214
Advances in Polymer Science

Advances in Polymer Science
Recently Published and Forthcoming Volumes

Photoresponsive Polymers II

Volume Editors: Seth R. Marder · Kwang-Sup Lee

With contributions by

C. Andraud · P. L. Baldeck · C. Barsu · H. Chermette
G. D. D'Ambruoso · R. Fortrie · H. Hoppe · D. V. McGrath
N. S. Sariciftci · O. Stéphan

 Springer

The series *Advances in Polymer Science* presents critical reviews of the present and future trends in polymer and biopolymer science including chemistry, physical chemistry, physics and material science. It is adressed to all scientists at universities and in industry who wish to keep abreast of advances in the topics covered.

As a rule, contributions are specially commissioned. The editors and publishers will, however, always be pleased to receive suggestions and supplementary information. Papers are accepted for *Advances in Polymer Science* in English.

In references *Advances in Polymer Science* is abbreviated *Adv Polym Sci* and is cited as a journal.

Springer WWW home page: springer.com
Visit the APS content at springerlink.com

ISBN 978-3-540-69452-6   e-ISBN 978-3-540-69454-0
DOI 10.1007/978-3-540-69454-0

Advances in Polymer Science ISSN 0065-3195

Library of Congress Control Number: 2008932858

Cover design: WMXDesign GmbH, Heidelberg
Typesetting and Production: le-tex publishing services oHG, Leipzig

Printed on acid-free paper

9 8 7 6 5 4 3 2 1 0

springer.com

ac

CCMR 10·31·2008

Advances in Polymer Science
Also Available Electronically

For all customers who have a standing order to Advances in Polymer Science, we offer the electronic version via SpringerLink free of charge. Please contact your librarian who can receive a password or free access to the full articles by registering at:

springerlink.com

If you do not have a subscription, you can still view the tables of contents of the volumes and the abstract of each article by going to the SpringerLink Homepage, clicking on "Browse by Online Libraries", then "Chemical Sciences", and finally choose Advances in Polymer Science.

You will find information about the

– Editorial Board
– Aims and Scope
– Instructions for Authors
– Sample Contribution

at springer.com using the search function.

Color figures are published in full color within the electronic version on SpringerLink.

Preface

Over the past 25 years or so there has been a revolution in the development of functional polymers. While many polymers as commodities represent huge markets, new materials with a high degree of functionality have been developed. Such specialty polymers play important roles in our day-to-day lives. The current volumes 213 and 214 of Advances in Polymer Science focus on photoresponsive polymers. In particular polymers that can either change the properties of a beam of light that passes through them or who change their properties in response to light. Volume 213 starts with an introduction to two-photon absorption by Rumi, Barlow, Wang, Perry, and Marder. In this chapter they develop the basic concepts of two-photon absorption, and describe structure–property relationships for a variety of symmetrical and unsymmetrical molecules. The applications of these materials in 3D microfabrication of polymers, metals, and oxide materials are detailed in the chapter entitled "Two-Photon Absorber and Two-Photon Induced Chemistry" contributed by the same group of authors. Then Belfield, Bondar, and Yao describe the molecules, dendrimers, oligomers, and polymers that can be excited by two-photon absorption and their application in processing materials with three-dimensional spatial control in their chapter entitled "Two-Photon Absorbing Photonic Materials." Specifically they describe the development of symmetrical and polar conjugated materials for two-photon absorption and their use as photo-initiators for 3D microfabrication. Juodkazis, Mizeikis, and Misawa also explore multiphoton processing of materials in their chapter, and provide more focus on the processing aspects of these materials and discuss the state-of-the-art in resolution.

In Volume 214, Hoppe and Sariciftci describe how organic semiconducting polymers can be used to produce electrical power when excited by light in the chapter entitled "Polymeric Photovoltaic Devices." In particular the authors review approaches based upon blends of conjugated polymers with small molecules that are approaching a point where they can be considered for commercialization. This is followed by a chapter by McGrath and D'Ambruoso entitled "Energy Harvesting in Synthetic Dendritic Materials" where they describe dendritic materials that can absorb light across various parts of the UV–visible spectrum and funnel energy down to a low energy absorber, which can be useful for a variety of applications including photovoltaics. Finally,

Baldeck and Andraud provide a chapter entitled "Exitonic Coupled Oligomers and Dendrimers for Two-Photon Absorption," wherein the concepts of excitonic coupling are developed and their relevance to multi-photon absorption processes are described.

The editors hope that these volumes will provide the reader with an overview of various aspects of photoresponsive polymers. We recommend that readers also examine other volumes in this series to learn more about related topics. In addition the editors thank the authors of the chapters in these volumes and the staff of Springer for their contribution to these volumes and accept responsibility for any errors or inaccuracies.

Atlanta & Daejeon, May 2008   S. R. Marder and K.-S. Lee

Contents

Contents of Volume 213

Photoresponsive Polymers I

Volume Editors: Marder, S. R., Lee, K.-S.
ISBN: 978-3-540-69448-9

Adv Polym Sci (2008) 214: 1–86
DOI 10.1007/12_2007_121
© Springer-Verlag Berlin Heidelberg
Published online: 17 October 2007

Polymer Solar Cells

Harald Hoppe[1] (✉) · N. Serdar Sariciftci[2]

[1]Institute of Physics, Experimental Physics I, Technical University of Ilmenau,
Weimarer Str. 32, 98693 Ilmenau, Germany
harald.hoppe@tu-ilmenau.de

[2]Linz Institute for Organic Solar Cells (LIOS), Physical Chemistry,
Johannes Kepler University Linz, Altenbergerstr. 69, 4040 Linz, Austria

Abstract Polymer solar cells, a highly innovative research area for the last decade until today, are currently maturing with respect to understanding of their fundamental processes of operation. The increasing interest of the scientific community is well reflected by the—every year—dynamically rising number of publications. This chapter presents an overview of the developments in organic photovoltaics employing conjugated polymers as active materials in the photoconversion process. Here the focus is on differentiating between the various material systems applied today: polymer–fullerene, polymer–polymer, polymer–nanoparticle hybrids, and polymer–carbon nanotube combinations are reviewed comprehensively.

1
Introduction

A polymer solar cell is defined by applying semiconducting conjugated polymers [1–3] as active components in the photocurrent generation and power conversion process within thin film photovoltaic devices that convert solar light into electrical energy. In the year 2000, Heeger, MacDiarmid, and Shirakawa received the Nobel Prize for Chemistry for the "discovery and development of conducting polymers", representing a new class of materials.

Conducting polymers generally exhibit an alternating single bond–double bond structure (conjugation) based on sp^2-hybridized carbon atoms. This leads to a highly delocalized π-electron system with large electronic polarizability. This enables both absorption within the visible light region, due to π–π^* transitions between the bonding and antibonding p_z orbitals, and electrical charge transport—two requirements that need to be met by semiconductors for power generation in solar cells. Using conjugated polymers to fabricate optoelectronic devices such as organic light-emitting diodes (OLEDs), organic field-effect transistors (OFETs), and organic solar cells (OSCs) is attractive because of their unique processability from solution [4]. Conjugated polymers, functionalized by solubilizing side-chain derivations, can be readily dissolved in common organic solvents—or even water—and thus can be used as "ink" for all kinds of deposition processes forming thin and homogeneous films. This property is especially interesting when combined with classical printing techniques, as it enables both spatially localized deposition (e.g., by inkjet or offset printing) and large area roll-to-roll manufacturing, allowing high-throughput production easily surmounting those achieved by classical semiconductor batch processing.

Charge carrier mobilities in organic semiconductors are generally much lower than those of their inorganic counterparts [5]. This disadvantage is partly balanced by high absorption coefficients [6, 7] and long-lived charge carriers [8–10], for example in polymer–fullerene blends. Furthermore, recent charge carrier mobilities obtained in polymer [11] and fullerene films [12], which are close to or even larger than those obtained in amorphous silicon films, make them an interesting alternative for, e.g., thin film transistor (TFT) arrays as used in liquid crystal (LCD) or OLED displays.

The structures of several conjugated polymers used in organic solar cells, along with a fullerene, are illustrated in Fig. 1. Three important and commonly used hole-conducting, donor-type polymers are MDMO-PPV (or OC_1C_{10}-PPV) (poly[2-methoxy-5-(3,7-dimethyloctyloxy)]-1,4-phenylenevinylene), P3HT (poly(3-hexylthiophene-2,5-diyl)), and PFB (poly[9,9′-dioctylfluorene-co-bis-N,N'-(4-butylphenyl)-bis-N,N'-phenyl-1,4-phenylenediamine]). Typical electron-conducting acceptors are the polymers CN-MEH-PPV (poly[2-methoxy-5-(2′-ethylhexyloxy)]-1,4-(1-cyanovinylene)-phenylene) and F8BT (poly(9,9′-dioctylfluorene-co-benzothiadiazole)), and a soluble derivative of C_{60}, called PCBM ([6,6]-phenyl C_{61}-butyric acid methyl ester). All of these materials are solution-processible due to side-chain solubilization and the polymers yield strong photo- and electroluminescence.

Conjugated polymers exhibit an alternating single bond–double bond structure of sp^2-hybridized carbon atoms. The electrons in the p_z orbitals of each sp^2-hybridized carbon atom form collectively the π band of the conjugated polymer. Due to the isomeric effects these π electrons are delocalized, resulting in high electronic polarizability. The Peierls instability splits the originally half-filled p_z "band" into two, the π and π^* bands. Upon light ab-

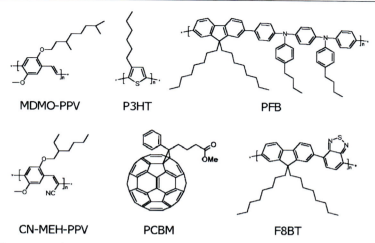

Fig. 1 Structures of conjugated polymers and a soluble C_{60} derivative commonly applied in polymer-based solar cells

sorption electrons may be excited from the bonding π into the antibonding π^* band. This absorption corresponds to the first optical excitation from the highest occupied molecular orbital (HOMO) to the lowest unoccupied molecular orbital (LUMO). The optical band gaps of most conjugated polymers are around 2 eV.

The portion of the solar light that typical polymeric solar cells absorb is limited. In Fig. 2a the absorption coefficients of thin films from two common conjugated polymers and PCBM are shown in comparison to the AM 1.5 solar

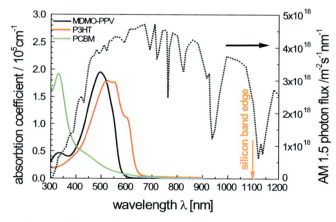

Fig. 2 Absorption coefficients of two conjugated polymers and a fullerene derivative PCBM, which represent the most often studied polymer–fullerene systems, are shown together with the AM 1.5 standard solar spectrum

spectrum. While the silicon band gap and onset of optical absorption spectrum is around 1.1 eV (ca. 1100 nm), most organic semiconducting polymers used today in photovoltaics utilize only the portion of the solar spectrum below 650 nm (larger than $\sim$2 eV). The absorption coefficients are comparatively high ($\sim$$10^5$ cm^{-1}) and allow for efficient absorption in very thin active layers.

During the last decade polymer solar cells have attracted a steadily increasing interest in both science and industry [7, 13–16]. The growing number of scientific publications within this field of research since 1990 impressively demonstrates this fact (Fig. 3) [17]. In fact this surge of interest has corresponded with the accelerating improvements in power conversion efficiency obtained during the last decade, currently reaching about 4–5% [18–21].

While in the early 1990s power conversion efficiencies in single layer, single component devices were still limited to less than 0.1% [22–25], improvements over the turn of the millennium are attributed to a great extent to the introduction of the donor–acceptor "bulk heterojunction" concept, which makes use of two electronic components that exhibit an energy offset in their molecular orbitals [26–34].

In this chapter we will briefly introduce the basic working principles of polymer solar cells, review the different device architectures (single layer, bilayer, and blend), and present an overview of the following most often studied material systems as applied within the photoactive layer: polymer–fullerene, polymer–polymer, polymer–nanoparticle (hybrid), and polymer–nanotube combinations. Table 1 displays an overview of current record efficiencies obtained for the different polymer solar cell device concepts discussed in this chapter.

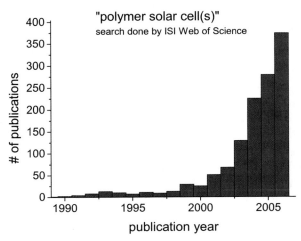

Fig. 3 Number of scientific publications contributing to the subject of "polymer solar cell(s)". Search done through ISI, Web of Science, 2007

Table 1 Record power conversion efficiencies and device parameters of polymer solar cells

System	Materials	Refs.	Short circuit current	Open circuit voltage	Fill factor	Power conversion efficiency
Polymer– fullerene	–	a	$9.35\,\mathrm{mA/cm^2}$ @ 100 mW	874 mV	64%	5.2%
Polymer– polymer	F8TBT:P3HT	[35]	$4\,\mathrm{mA/cm^2}$ @ 100 mW	1250 mV	45%	1.8%[b]
Polymer– hybrid	P3HT:CdSe nanorods	[36]	$8.79\,\mathrm{mA/cm^2}$ @ 92 mW	620 mV	50%	2.6%[b]
Polymer– nanotube	P3OT:SWCNT	[37]	$0.5\,\mathrm{mA/cm^2}$ @ 100 mW	750 mV	60%	0.22%

[a] Waldauf C (2007) Device ID: RM8; NREL certified, personal communication
[b] Corrected for spectral mismatch of the solar simulator

Donor–acceptor diblock copolymers constitute a further interesting class of materials based on the bulk heterojunction concept being developed lately [38–43].

1.1
Basic Working Principles of Polymer Solar Cells

Incident light that is absorbed within the photoactive layer of a polymer solar cell leads first to the creation of a bound electron–hole pair—the "exciton". These excitons diffuse during their lifetime with diffusion lengths generally limited to about 5–20 nm in organic materials [44–48]. This consideration is important to the design of active layer architectures. If an exciton does not eventually separate into its component electron and hole, it eventually recombines by emitting a photon or decaying via thermalization (nonradiative recombination). Hence, an exciton dissociation mechanism is required to separate the excitons which have binding energies ranging between 0.1 and 1 eV [49–53]. In single layer organic solar cells this may be achieved by the strong electric field present within the depletion region of a Schottky contact. Exciton dissociation in current polymer solar cells relies on gradients of the potential across a donor (D)/acceptor (A) interface, which results in the photoinduced charge transfer between these materials [26].

Upon light absorption in the donor an electron is excited from the HOMO into the LUMO. From this excited state the electron may be transferred into the LUMO of the acceptor. The driving force required for this charge transfer is the difference in ionization potential I_{D*} of the excited donor and the electron affinity E_A of the acceptor, minus the Coulomb correlations [26]. As

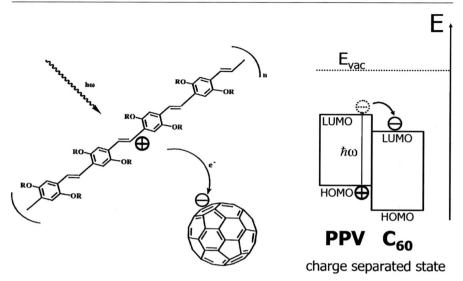

Fig. 4 Photoinduced charge transfer from a donor (here PPV) to an acceptor (here C_{60}) serves as a highly efficient charge separation mechanism in most polymer solar cells [26]

a result of the photoinduced charge transfer, the positively charged hole remains on the donor material whereas the electron is located on the acceptor. This is schematically depicted in Fig. 4 for a soluble derivative of poly(*para*-phenylenevinylene) as donor and C_{60} as acceptor.

This photoinduced charge transfer between conjugated polymers as donor and fullerenes as acceptor takes place within less than 50 fs [54]. Since all competing processes like photoluminescence ($\sim$ns) and back transfer and thus recombination of the charge ($\sim$μs) [8–10] take place on a much larger timescale, the charge separation process is highly efficient and metastable. These possible pathways for the decay of the system after excitation are displayed in Fig. 5 for comparison.

As a result, the photoinduced charge transfer is accompanied by a strong photoluminescence quenching of the otherwise highly luminescent conjugated polymer [26, 55]. In conjugated polymer–fullerene blends, the two signs of charge carriers resulting from exciton dissociation have been clearly identified by means of light-induced electron resonance (LESR) and photoinduced absorption (PIA) measurements [26].

Recently, geminate polaron pairs have been proposed for polymer–polymer [35, 56, 57] and polymer–fullerene [58, 59] blends as photoinduced intermediates. Here the hole and electron remain coulombically bound across the interface of the donor–acceptor heterojunction. Only via an electric field and/or a temperature-assisted secondary process, these geminate polaron pairs are dissociated, leading to free charge carriers. This can have a considerable effect on the achievable charge separation efficiencies, since the geminate

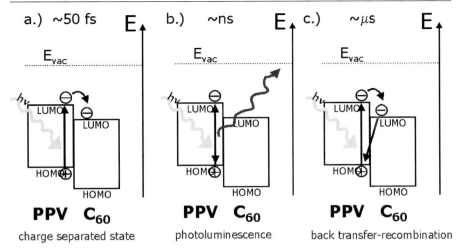

Fig. 5 Photoinduced processes in the donor–acceptor system. As the photoinduced charge transfer (**a**) occurs on a much smaller timescale than photoluminescence (**b**) and recombination (**c**), the charge separated state is efficiently formed and metastable

pair also decays without yielding free charges and this results in a significant loss channel for the photocurrent generation.

Once the charge carriers have been successfully separated, they need to be transported to the respective electrodes to provide an external direct current. Here the donor material serves to transport the holes whereas the electrons travel within the acceptor material. Thus, percolation paths for each type of charge carrier are required to ensure that the charge carriers will not experience the fate of recombination due to trapping in dead ends of isolated domains [60–62]. As such the bulk heterojunction has to consist of percolated, interpenetrating networks of the donor and acceptor phases.

When holes and electrons are separately transported within different spatial domains, the probability for charge recombination is reduced considerably leading to long charge carrier lifetimes. Charge carrier mobilities in conjugated polymer solar cells are around 10^{-4} cm^2/V s, thus long lifetimes are indeed required for extracting all photoexcited charge carriers from the photoactive layer. The charge carrier extraction is driven by internal electric fields across the photoactive layer caused by the different work function electrodes for holes and electrons.

The distance d that charge carriers can travel within the device is a product of charge carrier mobility μ, charge carrier lifetime τ, and the internal electric field F:

$$d = \mu \cdot \tau \cdot F \tag{1}$$

The internal electric field F that drives this drift current under photovoltaic operation generally originates from the difference in the electrode work functions. For the example of gold being the hole-accepting ($\Phi = 5.2$ eV) and aluminum being the electron-accepting electrode ($\Phi = 4.3$ eV), an internal electric field of 10^5 V/cm is given for an active layer thickness of 90 nm under short circuit conditions. Assuming charge carrier mobilities of 10^{-4} cm^2/V s and charge carrier lifetimes of 1 μs, a drift length $d = 10^{-4}$ cm = 100 nm at short circuit conditions is calculated.

In general the device function of thin organic solar cells, photodiodes, and even light-emitting diodes can be simplified using the metal–insulator–metal (MIM) model [63]. This is only valid when the organic semiconductors are not doped and as long as no significant space charge is built up during operation, which would result from unbalanced electron and hole transport.

The selectivity of charge injection/extraction into/from the molecular HOMO or LUMO levels ensures the rectifying diode behavior of these organic devices [64]. The different working regimes of these MIM devices due to externally applied voltages are shown in Fig. 6.

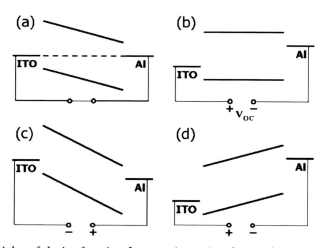

Fig. 6 Principles of device function for organic semiconducting layers sandwiched between two metallic electrodes: **a** short circuit condition, **b** flat band condition, **c** reverse bias, and **d** forward bias. Band bending effects at the ohmic contacts are neglected

Figure 6 represents different working regimes of a photovoltaic device, which correlate with points along a current–voltage (I–V) diagram as shown in Fig. 7: Fig. 7a corresponds to the short circuit photocurrent I_{SC}, and Fig. 7b to the flat band condition under open circuit voltage V_{OC}. In Fig. 7c the internal electric field is increased, corresponding to the condition in photodetectors or blocking behavior of diodes. For the case of forward bias (Fig. 7d), efficient charge carrier injection takes place and the direction of the current inside the device is reversed. This is the condition under which OLEDs are operating.

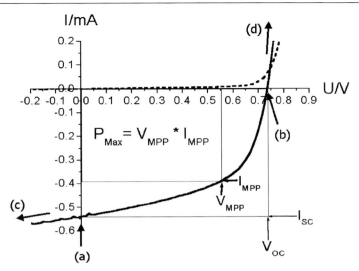

Fig. 7 Current–voltage characteristics of a polymer solar cell under illumination (*solid line*) and in the dark (*broken line*). The various situations (**a–d**) from Fig. 5 are shown for comparison

From Fig. 7 the calculation of the power conversion efficiency η can be derived: only the fourth quadrant of the I–V curve represents deliverable power from the device. One point on the curve, denoted as maximum power point (MPP), corresponds to the maximum of the product of photocurrent and voltage and therefore power. The ratio between $V_{MPP} \cdot I_{MPP}$ (or the maximum power) and $V_{OC} \cdot I_{SC}$ is called the fill factor (FF), and therefore the power output is written in the form: $P_{max} = V_{OC} \cdot I_{SC} \cdot FF$. Division of the output power by the incident light power res ults in the power conversion efficiency η:

$$\eta_{POWER} = \frac{P_{OUT}}{P_{IN}} = \frac{I_{MPP} V_{MPP}}{P_{IN}} = \frac{FF I_{SC} V_{OC}}{P_{IN}}. \tag{2}$$

As the transport of charges and thus the photocurrent is electric field dependent, close to the open circuit voltage the internal electric field is considerably reduced making the extraction of generated charge carriers less efficient, and leading to limitations in fill factor.

1.2
Device Architectures

The schematic design of a polymer solar cell is displayed in Fig. 8: the photoactive layer is usually sandwiched between an indium tin oxide (ITO)-covered substrate (glass or plastic) and a reflective aluminum back electrode. As the ITO substrate is transparent, illumination takes place from this side

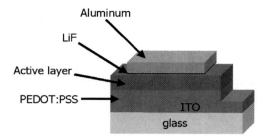

Fig. 8 Schematic design of an organic solar cell. The photoactive layer is sandwiched between optimized electron (Al) and hole extracting (ITO) electrodes

of the device. The two electrodes may be further modified by the introduction of a PEDOT:PSS (poly[3,4-(ethylenedioxy)thiophene]:poly(styrene sulfonate)) coating on the ITO side and a lithium fluoride (LiF) underlayer on the aluminum side, thereby improving the charge injection.

The device architecture of the photoactive layer has a strong impact on charge carrier separation and transport. For example in single layer (single material) devices, only photoexcitations generated close to the depletion region W of the Schottky contact may lead to separated charge carriers as a result of the limited exciton diffusion length. Therefore, only a small region denoted as the active zone contributes to photocurrent generation, as illustrated in Fig. 9.

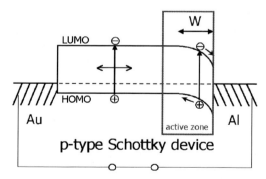

Fig. 9 In single layer single material devices, charge carriers can only be dissociated at the Schottky junction. Therefore only excitons generated close to the depletion region W can contribute to the photocurrent (denoted as the "active zone")

Bilayer devices [27, 65] apply the donor–acceptor concept introduced above: here the exciton is dissociated at their interface, leading to holes on the donor and electrons on the acceptor. Thus, the different types of charge carriers may travel independently within separate materials and bimolecular recombination is largely suppressed. Therefore light intensity-dependent photocurrent measurements in these systems exhibited a rather linear behavior of the photocurrent with respect to the light intensity, and monomolecular recombination processes dominate [27, 32, 46]. However, bilayer devices suffer also from an active zone limited by the exciton diffusion length, as only close to the geometrical heterojunction photoexcitations can lead to charge carrier generation, as indicated in Fig. 10.

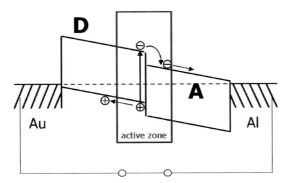

Fig. 10 In bilayer devices, charge carriers can be dissociated at the donor (D)–acceptor (A) material heterojunction. Only excitons generated within diffusion distance to the interface can contribute to the photocurrent

This limitation was finally overcome by the concept of the bulk heterojunction, where the donor and acceptor materials are intimately blended throughout the bulk [28–30]. In this way, excitons do not need to travel long distances to reach the donor/acceptor interface, and charge separation can take place throughout the whole depth of the photoactive layer. Thus the active zone extends throughout the volume, as illustrated in Fig. 11. Conse-

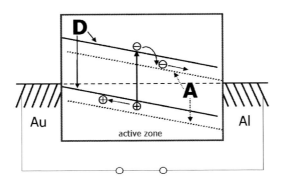

Fig. 11 In bulk heterojunction devices, charge carriers can be dissociated throughout the volume of the active layer. Thus every absorbed photon in the active layer can potentially contribute to the photocurrent

quently the bulk heterojunction concept led to major improvements of the photocurrent. Today, the bulk heterojunction serves as the state-of-the-art concept for polymer-based photovoltaics, leading to power conversion efficiencies of up to 5% [18–21].

Within the bulk heterojunction, the donor and acceptor domains are generally disordered in volume. For exciton dissociation and charge generation a fine nanoscale intermixing is required, whereas for the efficient transport of charge carriers percolation and a certain phase separation are needed to ensure undisturbed transport. Hence the optimization of the nanomorphology of the photoactive blend is a key issue for improving the efficiency of the photovoltaic operation [62, 66, 67].

Due to molecular diffusion of fullerenes at elevated temperatures, the phase separation may coarsen with time during operation in full sunlight, representing a morphological instability [68]. To overcome this degradation and for a better control of the nanomorphology itself, several concepts have been recently introduced to construct ordered bulk heterojunctions. They span a range from using self-assembled inorganic nanostructures for the infiltration of conjugated polymers [69, 70] up to self-organizing diblock copolymers [41, 43], where the two blocks carry the different functionalities of donor and acceptor, respectively. Figure 12 summarizes the discussed device architectures for comparison.

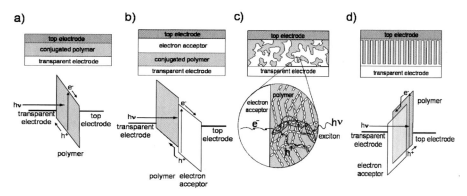

Fig. 12 Examples of device architectures of conjugated polymer-based photovoltaic cells: **a** single layer; **b** bilayer; **c** "disordered" bulk heterojunction; **d** ordered bulk heterojunction. (Reproduced with permission from [71], © 2005, American Chemical Society)

1.3
Influence of Electrical Contacts and Open Circuit Voltage

The proper choice of metallic contacts in OLEDs was shown to have a major influence on their performance [64]. This was further improved by adjusting the energy barrier height of the hole-injecting anode by interface modifications, like the application of polar self-assembled monolayers (SAMs) [72–75] or the introduction of interlayers using the highly doped polyelectrolyte PEDOT:PSS [76–79], and by application of lower work function metal cathodes and introduction of thin alkali metal salt interlayers (e.g., LiF) for reducing the electron injection barrier [80–87]. As a consequence, optimal electroluminescence quantum efficiencies are achieved when the two conditions, balanced charge carrier injection and balanced charge carrier mobilities, are met [88].

The development of suitable contacts for polymer solar cells has directly profited from the developments in light-emitting diodes, due to the injection/extraction similarity. Due to its lower barrier for hole transfer between most conjugated polymers and PEDOT:PSS as compared to ITO, this highly doped polymer electrode was applied at an early stage in solar cells as

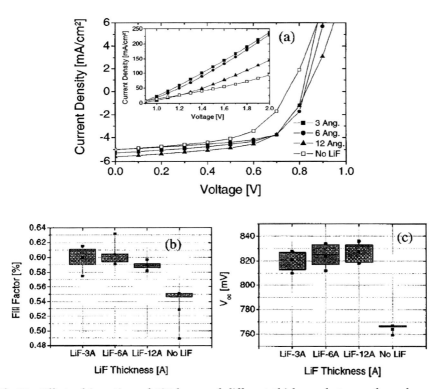

Fig. 13 Effect of insertion of LiF layers of different thickness between the polymer–fullerene blend and the aluminum electrode. The current–voltage characteristics indicate a more effective charge carrier injection (**a**), and as can be seen also from **b** and **c**, the fill factor as well as the open circuit voltage profit from LiF. (Reprinted with permission from [91], © 2002, American Institute of Physics)

well [13, 32, 89]. Further, the introduction of a LiF underlayer at the aluminum electrode brought about improvements in open circuit voltage and fill factor (Fig. 13) [13, 34, 90, 91].

Understanding the effect of this very thin—usually less than a nanometer and up to a few nanometers thick—LiF interlayer has been controversially debated. While it was observed that the insertion of the LiF underlayer resulted in an increased built-in potential [92], the mechanism for this effect is not fully clear. One possibility is that LiF leads to the formation of a thin barrier, forming a tunneling junction [80]. However, a thin SiO_2 interlayer behaved much differently from LiF [91]. As an additional advantage, the formation of trapping states due to chemical reactions between the aluminum and organic materials was prevented by the LiF interlayer [93–95].

Two major effects were proposed as causes for improved electron extraction: (a) upon sublimation of the subsequent aluminum layer the LiF dissociates, whereby metallic Li atoms may be formed that consequently n-dope the organic semiconductor (fullerene or polymer) under formation of Li^+ and, e.g., AlF_3 [86, 94, 96]; or (b) the LiF layer could result in an interfacial dipole layer shifting the work function of the electrode [82, 90, 91]. Both of these viewpoints have been shown to hold merit, as it was demonstrated by photoelectron spectroscopy that for very thin (sub-nanometer) layers of LiF dissociation and consequent n-type doping occurred, whereas for thicker layers (a few nanometers) the formation of a dipole was evidenced [97].

In general, the open circuit voltage (V_{OC}) of any solar cell is limited by the energy difference between the quasi-Fermi level splitting of the free charge carriers, i.e., the holes and the electrons [98], after their transport through the photoactive layer and the interfaces at the contacts. While for ideal (ohmic) contacts no energetic loss at the junction is expected, energy level offsets or band bending at non-ideal contacts will further reduce the V_{OC}. Recombination at the electrode may further reduce the quasi-Fermi level splitting.

The charge carriers require a net driving force toward the electrodes, which may in general result from internal electric fields and/or concentration gradients of the respective charge carrier species. The first leads to a field-induced drift and the other to a diffusion current. Without a detailed analysis one can generally assume that thin film polymer devices (< 100 nm) are mostly field drift dominated whereas thick devices, having effective screening of the electrical fields inside the bulk, are dominated by the diffusion of charge carriers, e.g., by concentration gradients created by the selective contacts.

The polaronic level of holes on the conjugated polymer donor phase is slightly above the HOMO of the polymer, and the transport level of the electrons is closely related to the LUMO level of the acceptor (n-type semiconductor, e.g., the fullerene). Thus, their resulting energetic splitting has to be related to the difference between the HOMO of the donor and the LUMO of the acceptor and conceptually determines the maximum open circuit potential of the photovoltaic device. This hypothesis has been proven by several studies reporting on the variation of the HOMO level of the donor polymer by using similar compounds with different oxidation potentials [99–102]. Figure 14 displays the linear relationship between the HOMO level of a larger set of conjugated polymers and the open circuit voltage applied in polymer–fullerene bulk heterojunctions.

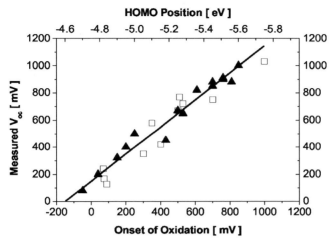

Fig. 14 Experimentally, a linear relationship between the HOMO level of the conjugated polymer (corresponds to onset of oxidation with respect to the Ag/AgCl reference electrode) and the measured open circuit voltage (V_{OC}) has been determined for a large number of donor polymers. (Reproduced from [102] with permission, © 2006, Wiley-VCH)

The same linear relationship was already earlier observed for the LUMO level of the acceptor fullerene by using fullerene derivatives with different first reduction potentials (see Fig. 15) [103]. This study has been recently extended to more fullerenes with smaller electron affinities, confirming this relationship [104].

The MIM model predicts the maximum V_{OC} being determined by the difference in the work functions of two asymmetrical electrodes, as long this is smaller than the effective band gap of the insulator [64]. Experimental data, however, showed strong deviations where the V_{OC} exceeded largely the expected difference between the electrode work functions [103]. Fermi level pinning between the fullerene and the metal electrode has been accounted for this. In another study, however, deviations from this pinning behavior have been found [105]. Thus, the individual energy level alignments between organic/metal interfaces are critical [72, 106–112]. Interfacial dipoles formed at the organic semiconductor/electrode interface change the effective metal work function and thus affect the V_{OC} as well [72, 108, 109, 113, 114].

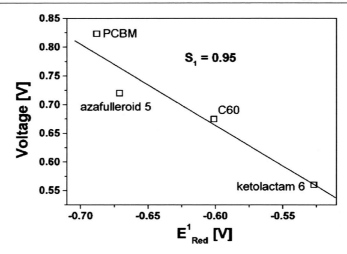

Fig. 15 Linear, close to unity dependence of the open circuit voltage on the LUMO level of the acceptor (first reduction potential E^1_{Red}, as determined electrochemically with respect to an Ag/AgCl reference electrode). (Reproduced from [103] with permission, © 2001, Wiley-VCH)

Another influence on V_{OC} may arise from aluminum electrodes, which can form a thin oxide layer at the interface to the organic materials [84, 115, 116], resulting in possible changes in the effective work function. A dependence of the V_{OC} on the nanomorphology arising from the fullerene content in bulk heterojunction blends was also observed [55, 117–121] and proposed to originate from the partial coverage of the cathode by fullerenes [118].

Furthermore, the theoretically expected dependencies of V_{OC} on temperature [122–124] and light intensity [28, 123–127] were also observed experimentally. This behavior needs to be correlated to the disorder broadening of the density of states of the organic semiconductors.

By varying the doping level of PEDOT:PSS electrochemically, it was demonstrated that the open circuit voltage indeed depends linearly on the work function of the hole-extracting electrode (see Fig. 16) [128]. Similar effects have been observed in another study of the metal electrode as well [105].

In conclusion, the open circuit voltage is limited primarily by molecular energy levels ($V_{OC} < $ LUMO$_{acceptor}$–HOMO$_{donor}$), with potential secondary limitations from the contacts (compare with Fig. 17), which themselves depend critically on the possible formation of interface dipoles, which can lead to substantial deviations from that simple relationship.

A major deviation from the above picture was found by Ramsdale et al., who investigated bilayer solar cells based on two polyfluorene derivatives [125]. When compared to the work function of both electrodes, the authors measured a V_{OC} exceeding those values by about 1 eV. This "overpotential" has been accounted for by a concentration driven diffusion current,

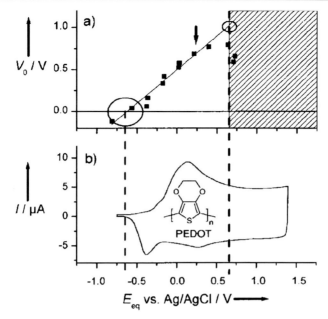

Fig. 16 Linear dependence of the compensation voltage V_0 (**a**), defined by the net photocurrent being zero, on the oxidation potential ($\approx$ work function) of electrochemically doped PEDOT layers (**b**) in polymer–fullerene solar cells. (Reproduced with permission from [128], © 2002, Wiley-VCH)

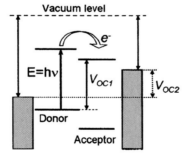

Fig. 17 Simple relationship of open circuit voltage V_{OC} for drift-current dominated bulk heterojunction polymer solar cells. The first limitation arises from the molecular energy levels (V_{OC1}); secondly, improper match with the contact work function may further reduce the achievable voltage to V_{OC2}. (Reprinted with permission from [105], © 2003, American Institute of Physics)

which needed to be balanced by a drift current due to change of external polarity. The particular situation is depicted in Fig. 18 and has been confirmed by Barker et al. by current–voltage modeling [125, 129].

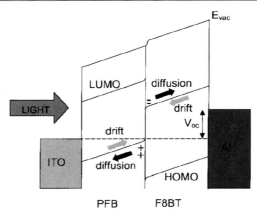

Fig. 18 Situation for a bilayer device, where the open circuit voltage exceeds the work function difference considerably by 1 V due to a concentration driven diffusion current. (Reprinted with permission from [125], © 2002, American Institute of Physics)

2
Polymer–Fullerene Solar Cells

Since the discovery of photoinduced charge transfer from a conjugated polymer (MEH-PPV) to a buckminsterfullerene (C_{60}) in 1992 by Sariciftci et al. [26], a dynamic development of solar cell devices exploiting this effect has followed. First applications of these two materials in bilayer geometry resulted in short circuit photocurrents following linearly the incident light intensity, even at higher illumination densities (Fig. 19) [27].

This linear dependence has been confirmed by Halls et al. using the same bilayer structure but employing PPV as the electron donor [44]. The authors estimated the exciton diffusion length of PPV to be in the range of 6–8 nm from both the spectral response and the absolute efficiency [44]. Later Roman et al. demonstrated optical modeling to be a useful tool for the optimization of such bilayer solar cells, which in their case was based on a polythiophene derivative and C_{60} [89]. The optical modeling was detailed by Petterson et al. [46].

In a next step of development a side group was attached to the C_{60} to allow for solution processing due to increased solubility in common organic solvents [130]. PCBM provides the best performances in polymer–fullerene solar cells, even to date. The first bulk heterojunction polymer solar cells were based on MEH-PPV and PCBM, and were presented by Yu et al. [29]. In these bulk heterojunctions, an intimate blending of the donor and acceptor components results in a very efficient exciton dissociation and thus charge carrier generation throughout the whole volume of the blend. MEH-PPV:PCBM blends with a mixing ratio of 1 : 4 spin-coated from *ortho*-1,2-

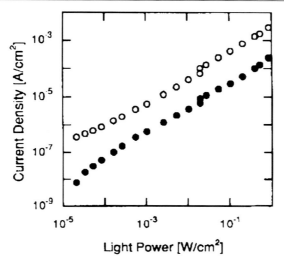

Fig. 19 Short circuit current (*closed circles*) and photocurrent at − 1 V bias (*open circles*) as a function of light intensity for the ITO/MEH-PPV/C$_{60}$/Au device. (Reprinted with permission from [27], © 1992, American Institute of Physics)

dichlorobenzene (ODCB) exhibited the best power conversion efficiencies and the authors found a nearly linear relationship between light intensity and photosensitivity [29].

Yang and Heeger investigated the nanomorphology of MEH-PPV:C$_{60}$ bulk heterojunctions spin-coated from ODCB by transmission electron microscopy (TEM) and revealed a bicontinuous phase structure of the two, which was finest for blending ratios of 1 : 1 [131]. Gao et al. also varied the blending ratio of MEH-PPV:C$_{60}$ bulk heterojunctions spin-coated from ODCB and found an enhancement of the photocurrent but a decrease of the photovoltage upon increasing the C$_{60}$ concentration in the blend [117]. While the improved photoresponse was related to the increase of total interfacial area at the heterojunction, the decrease in open circuit voltage was explained by the potential drop of the electron due to the transfer from polymer to fullerene LUMO. Moreover, Liu et al. investigated a broad range of organic solvents from which MEH-PPV bulk heterojunctions with up to 20 wt % of xylene-dissolved C$_{60}$ additions were spin coated. They clearly differentiated effects on photocurrent and open circuit voltage due to aromatic and nonaromatic solvents, respectively. Tetrahydrofuran (THF) and chloroform (nonaromatic) gave smaller currents but higher voltages than xylene, chlorobenzene, and dichlorobenzene (aromatic). The authors related this to a more intimate contact between the conjugated polymer backbone and the C$_{60}$ in the aromatic case. Additionally, a larger surface coverage by C$_{60}$—as shown by atomic force microscopy (AFM) in the phase imaging mode—was correlated with the reduction of V_{OC} for aromatic solvents (Fig. 20) [118].

Fig. 20 AFM micrographs (both height image and phase image) of MEH-PPV/C_{60} (20 wt %) composite films fabricated with **a** xylene, **b** DCB, and **c** THF. The phase image enables calculation of the C_{60} surface coverage. (Reproduced from [118] with permission, © 2001, Wiley-VCH)

Drees et al. developed a method to create diffuse bilayer heterojunctions between MEH-PPV and C_{60} by controlling the interdiffusion of a bilayer device by application of thermal annealing [132]. The MEH-PPV was first spin-coated from solution, and the C_{60} layer was thereafter thermally sublimed by a vacuum evaporation process. Annealing the devices at 150 and 250 °C led to an intensified increase in the photocurrents as compared to nonannealed devices. This has been interpreted as the result of diffusion-controlled formation of a larger interfacial contact area between MEH-PPV and C_{60} [132]. Using Auger spectroscopy in combination with ion beam milling, the existence of the diffuse interface due to annealing could be proven (see Fig. 21) [133].

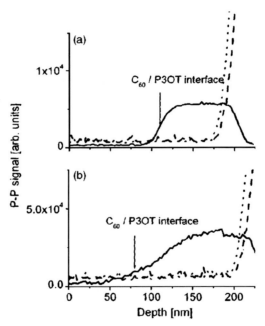

Fig. 21 Depth profiles of **a** an unheated P3OT/C_{60} bilayer device and **b** a P3OT/C_{60} bilayer device heated at 130 °C. The concentrations of sulfur (*solid lines*), indium (*dotted lines*), and oxygen (*dashed lines*) were monitored. The *arrow* indicates the position of the P3OT/C_{60} interface as determined from absorption measurements. The unheated bilayer shows a rather sharp interface between P3OT and C_{60}. For the interdiffused film, the P3OT concentration rises slowly throughout more than 100 nm (from 35 to 150 nm) of the bulk of the film. (Reprinted with permission from [133], © 2005, American Institute of Physics)

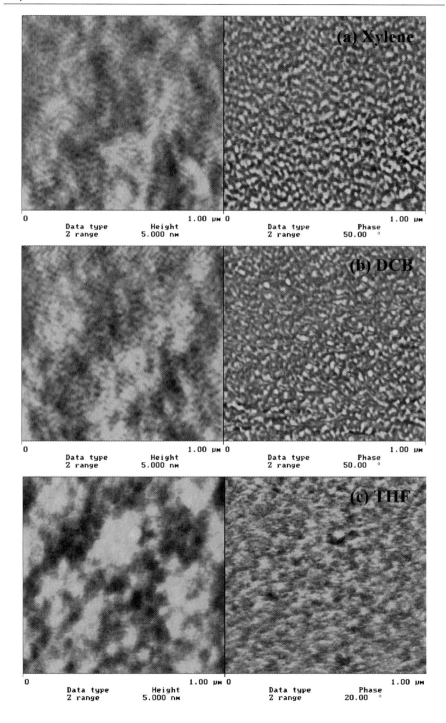

0   1.00 μм 0   1.00 μм

Data type   Height   Data type   Phase
Z range   5.000 nм   Z range   50.00 °

Data type   Height   Data type   Phase
Z range   5.000 nм   Z range   50.00 °

Data type   Height   Data type   Phase
Z range   5.000 nм   Z range   20.00 °

Shaheen et al. demonstrated that the solvent from which bulk heterojunctions of MDMO-PPV:PCBM were spin cast has a detrimental influence on the photocurrent generation and thus power conversion efficiency [34]. While the absorption of the films did not change, the spectral photocurrent exhibited an increase throughout the whole spectrum when spin cast from chlorobenzene instead of toluene. The authors related this to the formation of different nanomorphologies in the blend films [34]. Figure 22 displays the transmission and spectral photocurrent spectra, as well as the current–voltage curves, of either toluene or chlorobenzene cast MDMO-PPV:PCBM bulk heterojunctions. As can be seen, the photocurrent is increased throughout the whole spectrum where the solar cell absorbs light, leading to improvement by a factor of 2–3 for the short circuit photocurrent (I_{SC}).

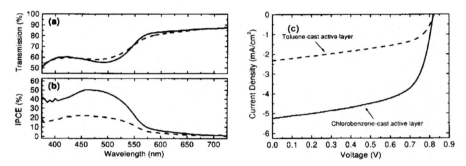

Fig. 22 Optical transmission spectra of 100-nm-thick MDMO-PPV:PCBM (1 : 4 by wt.) films spin cast onto glass substrates from either toluene (*dashed line*) or chlorobenzene (*solid line*) solutions (**a**). Incident photon to collected electron (IPCE) spectra (**b**) and current–voltage characteristics (**c**) for photovoltaic devices using these films as the active layer. (Reprinted with permission from [34], © 2001, American Institute of Physics)

The solvent-dependent efficiency increase of up to 2.5% by Shaheen et al. has motivated many following studies, aiming to discover the underlying nanomorphology and phase distribution of MDMO-PPV and PCBM [62]. Martens et al. applied TEM to investigate similar MDMO-PPV:PCBM bulk heterojunctions [134, 135]. While a 1 : 1 mixture of MDMO-PPV:PCBM spin cast from toluene appeared homogeneous in TEM, the authors observed the occurrence of stronger phase separation with increasing volume fraction of fullerene content. For blending ratios of 1 : 4 (MDMO-PPV:PCBM) they showed by cross-sectional TEM images that large, dark fullerene clusters are inside the film for toluene, whereas for chlorobenzene cast films the cluster size is considerably smaller. Chlorobenzene cast films were homogeneous up to blending ratios of 1 : 2, and the authors concluded that for higher blending ratios the PCBM clusters are surrounded by the very same homogeneous blend of 1 : 1 and 1 : 2 for toluene and chlorobenzene, respectively [135].

Furthermore, Martens et al. have shown by AFM that the drying time is an important parameter for the size of the phase-separated structures. By introducing a hot air flow over a drying film, the drying time could be decreased and consequently the extent of phase separation was reduced [136].

Hoppe et al. studied MDMO-PPV:PCBM solar cells for decoding the different nanophases within these MDMO-PPV:PCBM blends cast from the two solvents (toluene and chlorobenzene) [55]. A large difference in the scale of phase separation could be identified as a major difference between toluene and chlorobenzene cast blends (see Fig. 23), but this could not directly explain the observed differences in photocurrent generation [55].

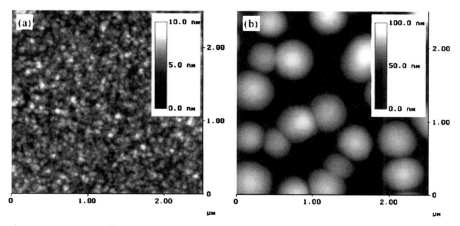

Fig. 23 Tapping mode AFM topography scans of MDMO-PPV:PCBM 1 : 4 blended films, spin cast from **a** chlorobenzene and **b** toluene solution. Clearly a larger scale of phase separation is observed for the toluene cast film. (Reproduced from [55] with permission, © 2004, Wiley-VCH)

Use of high-resolution scanning electron microscopy (SEM) allowed the uncovering of a further substructure in these polymer–fullerene blends, besides some larger fullerene clusters (see Fig. 24): MDMO-PPV nanospheres representing a coiled polymer conformation were detected together with some solvent-dependent amount of PCBM fullerenes [55, 60–62, 137].

The commonly observed larger scale of phase separation of the toluene cast MDMO-PPV:PCBM blends has generally been interpreted as the main reason for the reduced photocurrents in comparison to those of the chlorobenzene cast blends. It can then be expected that a lower charge carrier generation efficiency may result when exciton diffusion lengths of 10–20 nm are well exceeded by the PCBM cluster size (200–500 nm), since many excitons are generated within these clusters. Experimentally, it has been identified that indeed some unquenched photoexcitations give rise to residual PCBM photoluminescence in toluene cast blends, whereas in chlorobenzene cast

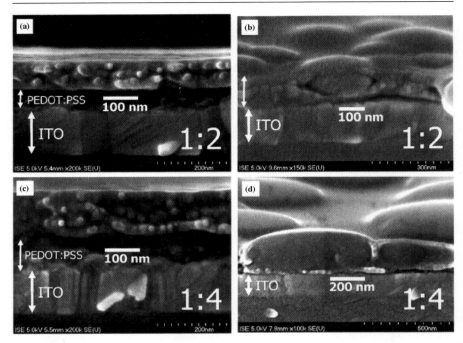

Fig. 24 SEM cross sections of chlorobenzene (**a**, **b**) and toluene (**c**, **d**) based MDMO-PPV:PCBM blends. Whereas chlorobenzene based blends are rather homogeneous, toluene cast blends reveal large PCBM clusters embedded in a polymer-rich matrix or skin layer. Small features—referred to as "nanospheres"—are visible in all cases and can be attributed to the polymer in a coiled conformation. The blending ratio is depicted in the lower right corner. (Reproduced from [55] with permission, © 2004, Wiley-VCH)

blends the fullerene photoluminescence could not be detected any more [55]. However, since the spectral photocurrent data show a vital contribution from the fullerene also in the case of toluene cast blends, the photocurrent in the region of the large fullerene cluster may have a significant contribution from dark triplet excitons that exhibit a longer lifetime and thus may diffuse longer distances—long enough to reach the heterojunction interface [62, 126].

Deeper insight can be gained from a scanning near-field optical microscopy (SNOM) study by McNeill et al., who resolved the local photocurrent obtained on MDMO-PPV:PCBM toluene cast blends [138]. The authors revealed that the photocurrent was considerably reduced on top of the small hills caused by the PCBM clusters (Fig. 25), whereas it stayed nearly constant over the surface of chlorobenzene cast blends [138].

As could already be inferred from the cross-sectional SEM images, the fullerene clusters are in fact surrounded by a polymer-rich skin layer [55, 60]. Using Kelvin probe force microscopy Hoppe et al. were able to confirm this by the detection of a considerably increased work function on top of the polymer

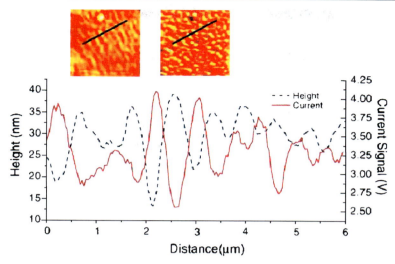

Fig. 25 Height and local photocurrent signal obtained by near-field scanning photocurrent measurements. At the top, both the topographic (*left*) and photocurrent (*right*) images are shown. (Reprinted from [138], © 2004, with permission from Elsevier)

embedded clusters [60]. The larger work function on top of the clusters as compared to the polymer-rich matrix around the clusters and on chlorobenzene based blends is a clear indication of an increased hole density at the film surface, which in turn points to the presence of the hole-conducting polymer [60]. This polymer-rich skin layer around the fullerene clusters in return represents a loss mechanism for the photocurrent, as electrons originating from the fullerene clusters will simply suffer recombination due to the high density of holes within the skin. Therefore the reduced photocurrents observed for toluene cast MDMO-PPV:PCBM films are caused by missing percolation of the fullerene phase toward the electron-extracting electrode. The situation for chlorobenzene and toluene cast blends is depicted schematically in Fig. 26 [61].

In conclusion, not only the observed larger scale of phase separation but also the difference in the material's phase percolation and thus charge transport properties influence the photovoltaic performance. As such, it becomes evident that the charge carrier mobility measured in these devices must be a function of the blend morphology [139–143]. Furthermore, the electron and hole carrier mobilities depend strongly on the polymer–fullerene blending ratio. Interestingly, the hole mobility of the donor polymer is increased considerably in blends with fullerenes (see Fig. 27) [142, 144–147].

Thus a high percentage of PCBM (80%) is needed in MDMO-PPV:PCBM solar cells for optimal photovoltaic performance [120]. For strongly unbalanced charge carrier mobilities between donor and acceptor, the buildup of space charge regions in the photoactive layer ultimately limits the photocur-

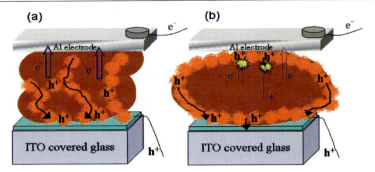

Fig. 26 Differences in the chlorobenzene (**a**) and toluene (**b**) based MDMO-PPV:PCBM blend film morphologies are shown schematically. In **a** both the polymer nanospheres and the fullerene phase offer percolated pathways for the transport of holes and electrons, respectively. In **b** electrons and holes suffer recombination, as the percolation is not sufficient. (Reprinted from [61], © 2005, with permission from Elsevier)

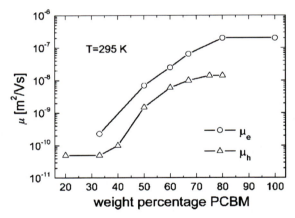

Fig. 27 Compositional dependence of electron and hole mobilities in MDMO-PPV:PCBM blend films as obtained by space charge limited diode currents. Clearly the mobility for holes is increased upon addition of the fullerene. (Reproduced from [144] with permission, © 2005, Wiley-VCH)

rent and the photovoltaic performance by a maximum possible fill factor of 42% [148].

Katz et al. investigated the performance of the polymer solar cells under elevated temperatures in the range of 25–60 °C, which represents real operating conditions due to heating under solar irradiation [122]. While the open circuit voltage (V_{OC}) decreased linearly with temperature, the short circuit current (I_{SC}) and the fill factor (FF) increased up to about 50 °C, followed by a saturation region (Fig. 28). These effects overcompensated the dropping V_{OC} and thus the efficiency was maximal for a 50 °C cell temperature [122].

to describe accurately the light intensity dependence of both the open circuit voltage and the short circuit photocurrent for polymer–fullerene solar cells [155, 156].

In contrast, Schilinsky et al. and Waldauf et al. used an extended numerical description according to the p–n junction model and demonstrated as well a proper description of light intensity-dependent device current–voltage characteristics [127, 157, 158]. Thus, several numerical models for the electrical description of polymer–fullerene photovoltaic devices have been presented in the literature to date, and there is an ongoing discussion in the scientific community about them.

For an adequate understanding of the photogeneration process in these thin film multilayer solar cells, optical modeling of light propagation and absorption is required. Due to the thin film thickness for most of the layers in the device, coherent light propagation needs to be considered to take possible interference effects properly into account. The commonly applied numerical description is done by the so-called transfer matrix formalism [159]. The resulting generation density of excitons can then be used as input for the electrical description of the device. Optical modeling has been done for both bilayers [46, 160, 161] and bulk heterojunction [6, 162–166] solar cells.

Synthesis via the "sulfinyl route" led to a reduced number of defects on the MDMO-PPV donor polymer and showed some improved performances in MDMO-PPV:PCBM bulk heterojunctions [167, 168]. The lower defect density resulted in a more regioregular (head-to-tail) order within the MDMO-PPV, leading to charge carrier mobility improvements and ultimately to an improved efficiency of 2.65% for MDMO-PPV:PCBM based bulk heterojunctions [169]. This was accompanied by a fill factor of 71% [169], which to date has not been exceeded by any other polymer solar cell device.

A further improvement of MDMO-PPV based bulk heterojunctions was achieved by the application of a new C_{70} fullerene derivative, which was substituted with the same side chains as PCBM and is therefore called [70]PCBM [170]. Due to the reduced symmetry of C_{70} as compared to the football sphere (icosahedral symmetry) of C_{60}, more optical transitions are allowed and thus the visible light absorption is considerably increased for [70]PCBM. This led to an improved external quantum efficiency (EQE) of MDMO-PPV based solar cells reaching up to 66% (Fig. 30). As a result the power conversion efficiency was boosted to 3% under AM 1.5 solar simulation at $1000\,W/m^2$ [170].

Thermally activated PCBM diffusion and formation of crystalline aggregates within blends with PPV derivatives were observed even at moderate temperatures [55, 68, 137]. In contrast, polythiophene based polymer-fullerene solar cells had an overall performance improvement upon thermal annealing steps [171, 172]. This improvement has been mainly correlated with an improved order in the film. This is especially true in the case of polythiophene, which is known to convert to a more ordered phase upon

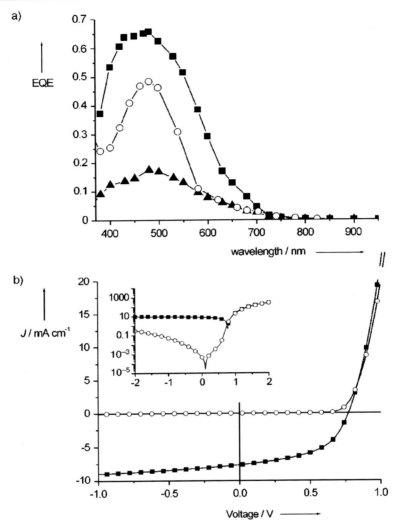

Fig. 30 Photovoltaic properties of an MDMO-PPV based polymer–fullerene solar cell with an active area of 0.1 cm². **a** External quantum efficiency (EQE) of [70]PCBM:MDMO-PPV cells, spin-coated from chlorobenzene (*triangles*) and ODCB (*squares*), and of [60]PCBM:MDMO-PPV devices spin-coated from chlorobenzene (*open circles*); **b** current–voltage characteristics of [70]PCBM:MDMO-PPV devices, spin-coated from ODCB in the dark (*open circles*) and under illumination (AM 1.5, 1000 W/m²; *squares*). The inset shows the *I–V* characteristics in a semilogarithmic plot. (Reproduced with permission from [170], © 2003, Wiley-VCH)

thermal annealing [174] or chloroform vapor treatment [175]. The ordered phase of P3HT is known to lead to high charge carrier mobilities of up to 0.1 cm²/V s [11].

Padinger et al. reported on postproduction treatments of P3HT:PCBM bulk heterojunction solar cells [172]. After a combined heat and dc voltage treatment, the power conversion efficiency could be boosted to 3.5%. Applying only the thermal annealing step itself raised the efficiency from 0.4 to 2.5%. However, the diode characteristics were further improved by application of the relatively strong forward dc current at 2.7 V. The authors explained the improved diode characteristics upon dc current application by the burnout of parasitic shunt currents. In Fig. 31 the effect of postproduction treatments on the I–V characteristics is presented. The absorption increase due to the annealing in similar devices was estimated to be around 40% and thus could not fully account for the improved device performance [163].

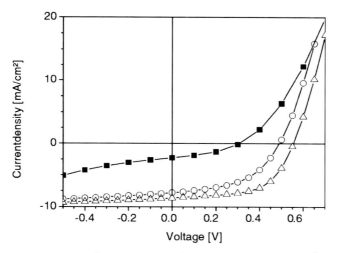

Fig. 31 I–V measurements of P3HT:PCBM solar cells under $80\,mW/cm^2$ AM 1.5 solar spectrum simulation (light). The photocurrent and the diode characteristics improved from untreated (U, *squares*) over thermal annealing (T, *open circles*) to thermal annealing in combination with the application of external voltage (T + I, *open triangles*). (Reproduced with permission from [172], © 2003, Wiley-VCH)

Chirvase et al. showed the effect of annealing on optical absorption to be rather correlated with a molecular diffusion of PCBM out of the polythiophene matrix [175]. Furthermore, the growth of PCBM clusters led to the formation of percolation paths and thus to improved photocurrents. Improved ordering of P3HT domains via interchain interaction [176] and a reduction of interface defects [177] has been previously connected to thermal annealing. Chirvase et al. showed that thermal annealing of pristine PCBM or P3HT films yielded only slight changes in the absorption, whereas annealing of blends resulted in a large increase of P3HT absorption (see Fig. 32) [175].

The growth of large micron-sized PCBM crystal domains depended on the initial concentration of PCBM in the blend as well as on the duration of the

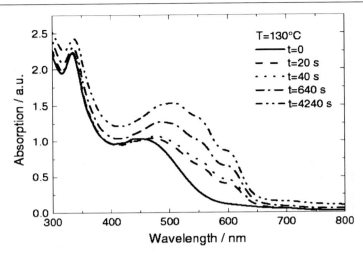

Fig. 32 Absorption spectra of a P3HT:PCBM composite film as cast (*solid curve*) and after four successive thermal annealing steps, as indicated in the legend. The PCBM concentration is 67%. (Reprinted with permission from [175], © 2004, Institute of Physics Publishing)

annealing process. Figure 33 shows the surface topography of P3HT:PCBM films without and with an aluminum electrode for two different PCBM concentrations (50 and 75%). Clearly a dendrite structure is observed with increasing size for increasing PCBM content.

Kim et al. suggested that vertical phase segregation between P3HT and PCBM results from the thermal annealing [178], where P3HT is segregated adjacent to the PEDOT:PSS electrode. Thus the holes could be transported more efficiently to the PEDOT:PSS electrode and electrons directly to the top aluminum contact, yielding better diode rectification [178]. In addition, the authors investigated the influence of the annealing temperature on the device parameters and found the best results for annealing at 140 °C.

Yang et al. reported TEM and corresponding electron diffraction results of P3HT:PCBM blends [179]. An increase in crystallinity for both P3HT and PCBM phases as well as a fibrillar P3HT morphology with extended length are developed due to annealing [179]. The authors concluded that the expanding of the crystalline domains results in improved charge transport and device performance. Figure 34 shows the TEM images together with the electron diffraction images for the untreated and the annealed P3HT:PCBM blend.

Here the P3HT backbone is oriented vertically to the P3HT fibrils, with the π–π stacking direction parallel to the fibril axis of the P3HT crystals [180]. Hence, there are better charge carrier mobilities resulting along the π–π stacking direction (long axis of fibrils).

Erb et al. investigated the crystalline structure of P3HT:PCBM bulk heterojunctions by grazing incidence X-ray diffraction (GID-XRD) [181], showing

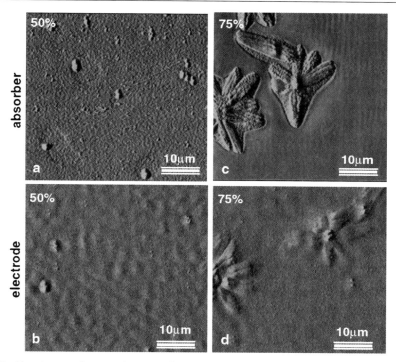

Fig. 33 Tapping mode AFM images of P3HT:PCBM films without (**a**, **b**) and with aluminum top electrode (**c**, **d**) at different PCBM concentrations. Large crystalline PCBM dendrites are observed for the larger fullerene concentration. (Reprinted with permission from [175], © 2004, Institute of Physics Publishing)

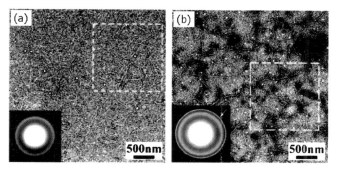

Fig. 34 TEM images in combination with electron diffraction of **a** untreated and **b** thermally annealed P3HT:PCBM blend films. (Reprinted with permission from [179], © 2005, American Chemical Society)

that upon annealing P3HT crystallites with a dimension of about 10 nm are grown. The polymer backbone orientation within these crystallites was found to be parallel to the substrate, with the side chains oriented perpendicular to the substrate (a-axis orientation of P3HT crystallites). The XRD signals before and after thermal annealing of P3HT:PCBM composites are shown in Fig. 35.

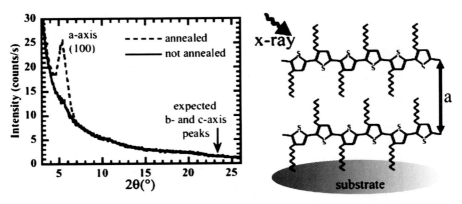

Fig. 35 Diffraction diagram (grazing incidence) of untreated and annealed P3HT:PCBM blend films deposited on glass/ITO/PEDOT:PSS substrates (*left*). The corresponding *a*-axis orientation of the crystals is shown on the *right*. (Reproduced with permission from [181], © 2005, Wiley-VCH)

The authors also correlated the increase in device efficiency with an increased crystallinity in the P3HT phase, but in contrast to Kim et al. they did not detect any PCBM crystals in the blend film [181]. Interestingly, the observed increase in optical absorption in the range of 1.9 to 3.0 eV upon annealing is directly correlated to the crystallization-induced ordering of P3HT in these blends (see with Fig. 36) [182].

It has been shown earlier by electron diffraction that PCBM is capable of self-organizing into crystalline order in pristine PCBM films [184]. In a more comprehensive study on MDMO-PPV:PCBM blends with varying composition, Yang et al. detected diffraction fringes in all films. The authors concluded that PCBM evolves in nanosized crystallites in the blend. Furthermore, upon annealing, the PCBM will organize into larger crystals—thereby destroying the original blend morphology. A conclusion of morphological instability at elevated temperatures was drawn [68]. In contrast to these observations, Yang et al. found a remarkable morphological stability over 1000 h at elevated temperatures of 70 °C for accelerated aging under illumination for the P3HT:PCBM blends [179]. Hence, they suggested that the ability of P3HT to crystallize has a stabilizing effect on the blend morphology. Schuller et al. reported similar results for an even higher temperature of 85 °C under simultaneous illumination with half a sun intensity at short circuit conditions [185]. Drees et al. demonstrated that by introducing an epoxy group into PCBM, the fullerenes could be linked together and polymerized by application of a catalyst in combination with annealing [186]. Thus, in comparison with conventional P3HT:PCBM 1 : 2

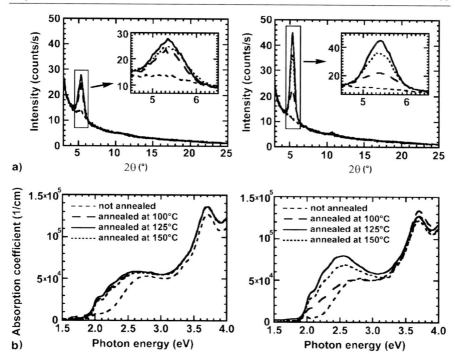

Fig. 36 XRD crystallinity (**a**) and absorption coefficient (**b**) of P3HT:PCBM blends spin cast from either chlorobenzene (*left*) or chloroform (*right*) at various annealing temperatures. Interestingly, the increase in crystallinity is accompanied by a quantitatively correlated increase in optical absorption in the 2.0–2.5 eV region. (Reprinted from [183], © 2006, with permission from Elsevier)

blends a morphological stabilization could be achieved. Figure 37 shows some tapping mode AFM images obtained on P3HT:PCBM and P3HT:PCBG blend films, both untreated and after thermal stress (annealing). Clearly the phase separation inside the PCBG containing blend is stopped before larger fullerene crystallites (Fig. 37c) can be developed.

Also Sivully et al. stabilized the P3HT:PCBM thin film nanomorphology, but in this case via application of amphiphilic diblock copolymers [187]. Reducing the PCBM concentration in blends with P3HT further down to less than 45%, Ma et al. confirmed that even at temperatures as high as 150 °C the film morphology was not considerably changed after 2 h and the overgrowth of fullerene aggregates was suppressed [18]. Thus, several encouraging results have been obtained so far toward long-term stable polymer–fullerene solar cells.

Recently there have been several reports published on 4–5% record efficiencies obtained on the P3HT:PCBM bulk heterojunction solar cell [18–21, 165, 188–190]. Ma et al. showed that annealing temperatures of 150 °C are required to obtain the highest efficiencies [18]. The authors assigned part

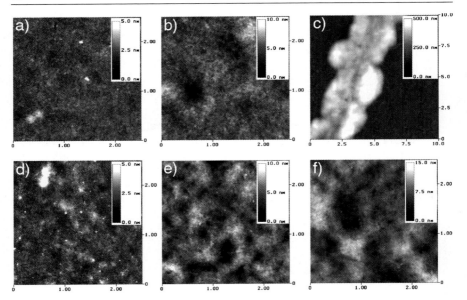

Fig. 37 AFM images of P3HT–PCBM (**a–c**) and P3HT–PCBG (**d–f**) blend films spin cast on glass. The images show unheated films (**a** and **d**), films annealed for 4 min at 140 °C (**b** and **e**), and films annealed for 1 h at 140 °C (**c** and **f**) [186]. (Reproduced by permission of The Royal Society of Chemistry)

of the performance improvement to the improved transfer of charges from the blend to the aluminum electrode. Huang et al. showed by time-of-flight (TOF) mobility measurements that the optimal composition between P3HT and PCBM has to be around 1 : 1 to obtain a balanced charge transport in the photovoltaic device (Fig. 38) [189].

Li et al. optimized the annealing temperature and the film thickness of P3HT:PCBM solar cells. Here only a 63-nm active layer thickness and temperatures of 110 °C resulted in optimal performances of around 4% [191]. For a larger film thickness, the photocurrent was reduced again [191]. By optical modeling based on the calculation of coherent light propagation, the thickness dependence of the solar cells can be understood by interference effects and allows for direct optimization of the active layer thickness (Fig. 39) [165, 192]. Note that there are regions, where an increase of the active layer thickness decreases the achievable photocurrent and *vice versa* [165].

Li et al. also showed that the use of ODCB as spin casting solvent in combination with slow drying of the active layer blend leads to high efficiencies for relatively thick film devices [21]. The authors accounted for the improved performance a self-ordering process of the polymer within the blend, resulting in balanced charge transport and a high fill factor [21]. Reyes-Reyes et al. reported high-efficiency solar cells based on P3HT:PCBM with a ratio of 1 : 0.8; here a rather short annealing step at 155 °C for 5 min was found

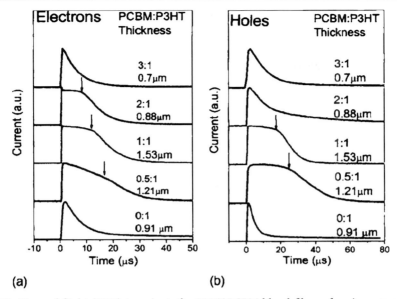

Fig. 38 Time-of-flight (TOF) transients for P3HT:PCBM blend films of various compositions. It can be clearly seen that the blending ratio of 1 : 1 yields nondispersive charge transport and similar mobilities (transition times) for both the electrons and the holes. (Reprinted with permission from [189], © 2005, American Institute of Physics)

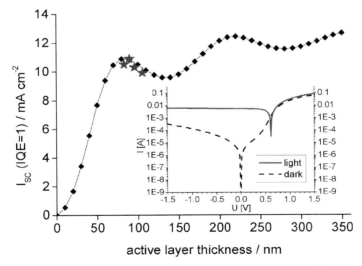

Fig. 39 Maximum theoretical short circuit currents from optical modeling (*diamonds*) and some from experiment (*stars*) are shown together for comparison. Indeed the experimentally determined photocurrents closely follow the theoretical prediction in the range investigated. The inset shows a corresponding *I–V* curve for the best device. (Reproduced with permission from [165], © 2007, Wiley-VCH)

to lead to the highest performances [19]. The authors attributed part of the success to a changed morphology as shown by AFM measurements. In a correlated study, the authors improved the annealing conditions for thicker active layers, thereby reducing the fullerene content even further down to about 38% [20]. Using high-resolution TEM, they detected small PCBM crystallites having sizes between 10 and 20 nm. The authors concluded that the individual crystallization of both materials requires a fine adjustment of the thermal treatment for optimal results [20].

Further, Kim et al. have demonstrated that the regioregularity of P3HT has a major influence on the order of the polymer phase and thus charge transport (Fig. 40) [190]. Only slightly lower regioregularity clearly reduced the crystallinity and charge carrier mobility in the polymer–fullerene blends. These effects have been observed for variations in the regioregularity as small as 4.5%, ranging from about 90 up to 95% [190].

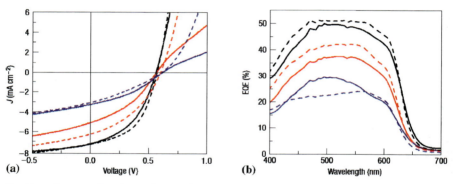

Fig. 40 Influence of regioregularity of P3HT in blends with PCBM on I–V characteristics and EQE. *Full lines* represent devices before annealing, *broken lines* after thermal annealing at 140 °C for 2 h. It is evident that larger regioregularity (*black*, 95.2%; *red*, 93%; and *blue* 90.7%) leads to larger photovoltaic performances. (Reprinted by permission from Macmillan Publishers Ltd: Nature Materials [190], © 2006)

Finally, the polymer molecular weight and/or polydispersity influence on the power conversion efficiencies: Schilinsky et al. found that higher weight fractions collected from the same P3HT batch show better performances than the lower weight fractions [193]. This observation is in agreement with measurements on field-effect transistors that showed higher mobilities for larger molecular weights of P3HT. Figure 41 summarizes the photovoltaic parameters obtained for P3HT:PCBM devices using the different weight fractions.

With the prospect of long-term stability [185, 188] and the ability to print polymer solar cells [188] with power conversion efficiencies of 4–5%, EQEs of over 75%, internal quantum efficiencies approaching unity [194], and fill factors of almost 70% [18, 21], the P3HT:PCBM system is at the moment highly optimized. The main limitations in reaching larger power conversion efficien-

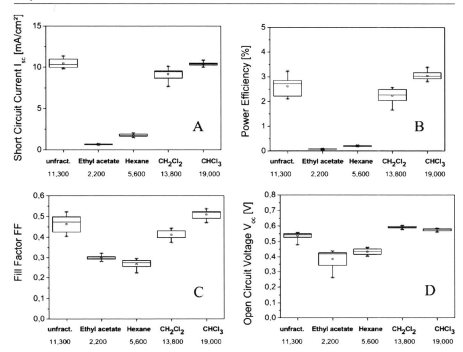

Fig. 41 Short circuit current (**A**), power conversion efficiency (**B**), fill factor (**C**), and open circuit voltage (**D**) of a number of P3HT:PCBM devices using different weight fractions (*x*-axis) as compared to the unfractionated sample. Clearly the larger weight fractions yield better values for all parameters investigated. (Reprinted with permission from [193], © 2005, American Chemical Society)

cies are found in the narrow absorption spectrum leading to an unsatisfactory overlap with the solar emission spectrum and the large potential losses due to the low LUMO level of the fullerene PCBM with regard to the P3HT LUMO level, resulting in rather large losses in short circuit current and open circuit voltage, respectively.

Polyfluorene derivatives have already been shown to exhibit much larger open circuit voltages ($V_{OC} > 1$ V) with PCBM than polythiophenes [195–197]. Comparable power conversion efficiencies of more than 4% have also been reported recently for high molecular weight fractions of PF10TBT (poly-[9,9-didecanefluorene-*alt*-(bis-thienylene)benzothiadiazole]), a polyfluorene based copolymer, thus demonstrating the potential for efficient high open circuit voltage polymer–fullerene solar cells [166].

However, in order to make progress toward 10% power conversion efficiency [102] further efforts need to be applied. One very promising concept is synthesizing "low band gap" conjugated polymers for solar energy conversion [198, 199]. Thereby a factor of 2 is expected on the photocurrent, if charge transport properties remain the same. Energetically it appears mostly favorable to lower the LUMO level of the polymer and thus reduce the band gap to allow for photon harvesting in the long wavelength range, while maintaining the open circuit voltage due a similar deep HOMO level. This image corresponds to the "ideal donor polymer" electronic properties [200].

One of the first employed low band gap materials in polymer–fullerene solar cells was PTPTB [201–203], extending the absorption spectrum to about 750 nm and yielding 1% power conversion efficiency [203]. Colladet et al. extended the absorption spectrum down to 800 nm based on thienylene–PPV derivatives [204]. Using a polyfluorene based copolymer, Zhou et al. achieved power conversion efficiencies of about 2.2%, with a spectral absorption range comparable to that of P3HT [205]. Campos et al. achieved sensitization of solar energy conversion down to 900 nm [206]. So far the largest range of absorption up to 1000 nm deep in the infrared region was demonstrated

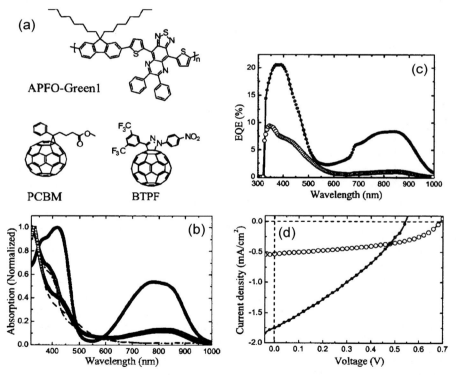

Fig. 42 Chemical structure (**a**) and absorption spectrum (**b**) of APFO-Green1, PCBM, and BTPF. The EQE (**c**) and *I–V* characteristics (**d**) under 100 mW/cm² solar spectrum simulation of APFO-Green1:PCBM 1 : 4 (*filled circles*) and APFO-Green1:BTPF 1 : 4 (*open circles*) photovoltaic devices are also shown. (Reprinted with permission from [207], © 2004, American Institute of Physics)

by Wang et al. and Perzon et al, using a fluorene based copolymer called APFO-Green1 [207–211]. Figure 42 displays the chemical structure and the absorption spectra of the polymer APFO-Green1 and of the fullerenes applied. The measured EQEs of the corresponding solar cells and their I–V characteristics are shown as well.

In combination with a C_{70} fullerene derivative—PTPF70—the system yielded improved power conversion efficiencies of 0.7% due the increased absorption of the C_{70} derivative [208, 210]. Using similar polyfluorene derivatives, power conversion efficiencies of about 0.9% (APFO-Green2) [212] and 2.2% (APFO-Green5) [213] were achieved. Other low band gap polymers, with absorption spectra extending up to 1100 nm, yielded efficiencies of around 1% [214–217].

Recently, the Konarka group achieved power conversion efficiencies of 5.2% for a low band gap polymer–fullerene bulk heterojunction solar cell, as confirmed by NREL (National Renewable Energy Laboratory, USA).[1] This encourages the practical use of this concept for low cost, large area production of photovoltaic devices.

Polymer–fullerene solar cells represent the most widely studied concept of polymer-molecule blend solar cells to date. Examples for the application of other acceptor dye molecules can be found here [218–220].

3
Polymer–Polymer Solar Cells

Polymer–polymer solar cells employ two different polymers as donor and acceptor components in the photoactive layer. These two polymers require a molecular energy level offset between their HOMO and LUMO levels to enable a photoinduced charge transfer. Due to the close vicinity of the respective molecular energy levels, polymer–polymer solar cells allow high open circuit voltages to be reached.

The first realizations of polymer–polymer bulk heterojunction solar cells were independently reported in the mid-1990s by Yu and Heeger as well as by Halls et al. [28, 30]. These solar cells were prepared from blends of two poly(*para*-phenylenevinylene) (PPV) derivatives: the well-known MEH-PPV (poly[2-methoxy-5-(2′-ethylhexyloxy)-1,4-phenylenevinylene]) was used as donor component, while cyano-PPV (CN-PPV) served as acceptor component (identical to MEH-PPV with an additional cyano (– CN) substitution at the vinylene group). The blends showed increased photocurrent and power conversion efficiency (20–100 times) when compared to the respective single component solar cells.

---

[1] Waldauf C (2007) Device ID: RM8-2; $I_{SC} = 9.346$ mA/cm^2, $V_{OC} = 0.8743$ V, FF = 63.81%, efficiency = 5.21%; NREL certified, personal communication.

The photosensitivity further increased upon application of a reverse bias (see Fig. 43, right), suggesting an application as photodetector as well. A sublinear relationship ($I_{SC} \sim I_{light}^{0.86}$) between the short circuit current I_{SC} and the incident light intensity I_{light} has been measured, indicating some bimolecular charge carrier recombination [28]. While charge transfer was the basis for the efficient operation of the MEH-PPV:CN-PPV blend solar cells, Halls et al. could show that similar blends of CN-PPV with bare PPV or DMOS-PPV (containing a silicon atom between the phenyl group and the alkyl side chain) resulted in energy transfer rather than charge transfer, thus enabling efficient OLED operation of these blends [221]. The authors related this behavior to the fact that the lowest excited state for the DMOS-PPV:CN-PPV system was an *intramolecular* CN-PPV transition, while for MEH-PPV:CN-PPV the smallest excited state was found to be *intermolecular* [221].

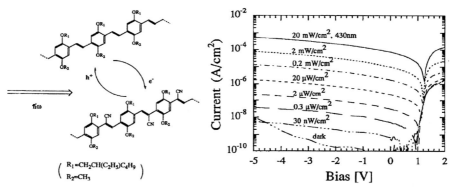

Fig. 43 Chemical structure of MEH-PPV and CN-PPV (*left*) as well as the monochromatic (430 nm) light intensity dependence of the *I–V* characteristics of MEH-PPV:CN-PPV blend polymer solar cells (*right*). (Reprinted with permission from [28], © 1995, American Institute of Physics)

Similar bulk heterojunctions with CN-PPV as acceptor polymer were realized with poly(3-hexylthiophene) (P3HT or denoted as PAT6 here) and PDPATPSi as donors, leading to a comparable behavior [31].

Granström et al. presented polymer–polymer solar cells using a regioregular phenyloctyl-substituted polythiophene derivative (denoted as POPT) as donor with (MEH-)CN-PPV as acceptor [32]. Here two different layers consisting of donor-rich (POPT:MEH-CN-PPV 19 : 1) and acceptor-rich (POPT:MEH-CN-PPV 1 : 19) blends were mechanically laminated to each other at an elevated temperature. As a result, a graded donor–acceptor heterojunction (diffuse bilayers) was fabricated, enabling both efficient exciton dissociation and selective charge transport to the respective electrodes [32]. The authors reported high EQEs and power conversion efficiencies, which were in part due to the absence of bimolecular recombination for this device

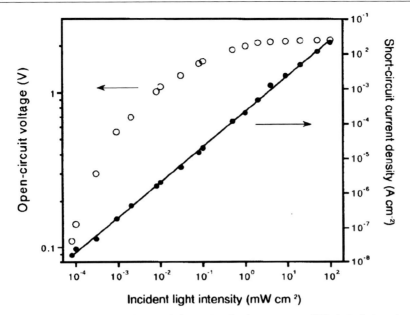

Fig. 44 Light intensity dependence of short circuit photocurrent (*filled circles*) and open circuit voltage of laminated POPT:(MEH-)CN-PPV diffuse bilayer polymer solar cells. The scaling factor of the current calculates as 1.02. (Reprinted with permission from [32], © 1998, Macmillan Publishers Ltd)

architecture, as inferred from the linear power dependency between I_{SC} and I_{light} ($I_{SC} \sim I_{light}^{1.02}$) (compare with Fig. 44).

The application of polymer precursors, resulting in insoluble PPV and BBL (poly(benzimidazo-benzophenanthroline)) ladder polymers enabled the fabrication of very efficient bilayer polymer solar cells, reaching 49% [33] and even 62% EQE (see Fig. 45) [222].

In these cases bimolecular recombination limited device efficiencies at higher light intensities; nonetheless, up to 1.1% power conversion efficiency was reached under full AM 1.5 solar irradiation [222]. Interestingly, the authors observed an open circuit voltage exceeding the work function difference of the respective electrodes by more than a factor of 2 for various acceptor polymers. The origin of this will be discussed on the basis of polyfluorene based polymer–polymer solar cells later in this section.

With molecular structures similar to the MEH-PPV:CN-PPV system, the intensively studied M3EH-PPV:CN-ether-PPV system—either as a blend or as a bilayer—resulted more recently in higher efficiencies under full AM 1.5 illumination ($100\,mW/cm^2$) (Fig. 46) [35, 223–225]. The first blend devices incorporated either a flat sintered titanium dioxide (TiO_2) or a PEDOT:PSS interlayer at the ITO interface. Blend devices with PEDOT:PSS and Ca electrodes led to power conversion efficiencies of 1% and EQEs of up to 23%.

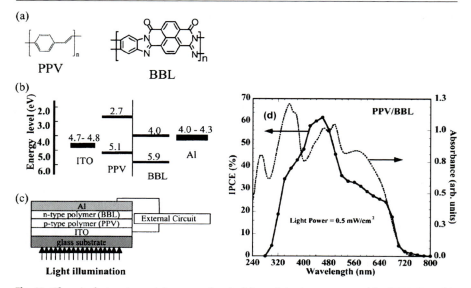

Fig. 45 Chemical structures (**a**), energy levels (**b**), and device structure (**c**) of PPV/BBL bilayers. High EQE (or IPCE) is shown together with the absorption spectrum of PPV/BBL bilayer solar cells. (*Left*: Reprinted with permission from [28], © 2000, American Institute of Physics; *Right*: Reprinted with permission from [193], © 2004, American Chemical Society)

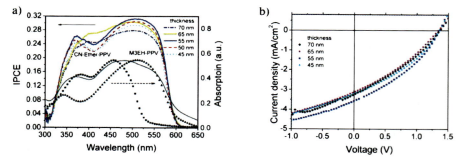

Fig. 46 Characteristics of solar cells based on M3EH-PPV:CN-ether-PPV blends (1 : 1 ratio by mass) for different thicknesses of the active layer after annealing at 110 °C. **a** Comparison of the IPCE and the absorption spectra of CN-ether-PPV (*squares*) and M3EH-PPV (*triangles*) of the blend layer (*black line*). **b** I–V characteristics under white light illumination at intensity of 100 mW/cm^2. (Reprinted with permission from [35], © 2005, American Chemical Society)

Breeze et al. found that the short circuit current increased as the blend layer thickness decreased, indicating the devices to be limited by charge transport losses [223]. Further improvements were achieved by Kietzke et al. using thermal annealing of blend layers reaching EQEs of 31% and power conversion efficiencies of 1.7%, which to date is one of the best efficiencies observed for polymer–polymer solar cells [226]. The authors attributed the improvement to a better ordering of the M3EH-PPV phase within the blend film.

These devices exhibited a linear scaling of the short circuit current with light intensity, indicating charge carrier recombination to be monomolecular. The absence of bimolecular recombination was in part related to the formation of a vertical phase segregation directed by the much lower solubility of M3EH-PPV as compared to CN-ether-PPV, leading to selective charge transport toward the electrodes. To explain the relatively low EQE—keeping a 95% efficient exciton quenching in mind—the authors proposed that after the exciton dissociation, no free charge carriers but bound polaron pairs (geminate pairs) are formed at the heterojunction. In a second step these may either dissociate or recombine to a lower lying exciplex state, leading to longer wavelength luminescent recombination [35]. Chasteen et al. verified the existence of an exciplex state by steady-state (see Fig. 47) and time-resolved photoluminescence measurements [224].

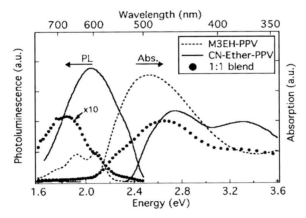

Fig. 47 Relative photoluminescence and absorption for M3EH-PPV, CN-ether-PPV, and blended films on quartz substrates. The higher-energy states in the neat films are highly quenched in the heterostructures, leaving exciplex emission at 1.8 eV. Relative photoluminescence data were excited at 2.82 eV (600 nm) and were corrected for optical density of the film. (Reprinted with permission from [224], © 2006, American Institute of Physics)

Kietzke et al. have shown for bilayer solar cells based on M3EH-PPV and several acceptor polymers with varying electron affinities and the fullerene derivative PCBM that the open circuit voltage is linearly related to the respective LUMO levels [225]. While CN-PPV-PPE acceptors resulted in an increased open circuit voltage of about 1.5 V, the fill factor and photocurrent were smaller than those for CN-ether-PPV [225].

Exciplex formation and subsequent transfer to a lower lying triplet state of MDMO-PPV in blends with PCNEPV (poly[oxa-1,4-phenylene-(1-cyano-1,2-vinylene)-(2-methoxy-5-(3,7-dimethyloctyloxy)-1,4-phenylene)-1,1-(2-cyanovinylene)-1,4-phenylene]) were identified as efficiency-limiting processes for photovoltaic devices [227–230]. Veenstra et al. demonstrated power conversion efficiencies of 0.75% by thermal treatment of the MDMO-PPV:PCNEPV active layer, by which the nanomorphology was altered [227]. For a set of PCNEPVs with three different molecular weights, the annealing temperature, under which the photocurrent was improved, showed strong correlation to the respective glass transition temperature of each weight fraction [227]. Quist et al. showed that annealing led to the formation of PCNEPV-rich domains, enabling the photoinduced electrons to diffuse away from the heterojunction interface and thus to escape from geminate recombination [228]. The authors further pointed out that low charge carrier mobility in combination with restricted charge collection at the electrodes were the limiting factors for solar cell performance.

Offermans et al. used several spectroscopic methods to study the formation and luminescence decay of exciplexes as a function of photoexcitation (of either polymer), and subsequent charge transfer at the heterojunction [229]. Exciplex formation was followed by a relaxation to the lower lying triplet state T_1 of the MDMO-PPV. The energetic scheme is depicted in Fig. 48.

While internal electric fields may offer a route to charge separation out of the exciplex, the free charge carriers may then recombine more easily to the triplet state due to the loss of spin correlation. This process resulted in electric field enhanced triplet formation and was identified as the major loss mechanism for the photovoltaic performance of MDMO-PPV:PCNEPV bulk heterojunctions [229, 230]. Using a set of several electron-accepting polymers, among them PCNEPV, Veldman et al. demonstrated that the open circuit voltage is again linearly related to the LUMO level of the acceptor. Since the charge separated state is lower than the singlet excited state and higher than the triplet state, this opens up the route to triplet-mediated recombination at the interface with MDMO-PPV [230].

Whereas few material systems studied yielding up to 1.5% power conversion efficiency are unique [231–234], blends and bilayers of polyfluorene based copolymers were most often investigated. PFB (poly[9,9'-dioctylfluorene-co-bis-N,N'-(4-butylphenyl)-bis-N,N'-phenyl-1,4-phenylenediamine]) is commonly used as the donor and F8BT (poly(9,9'-dioctylfluorene-co-benzothiadiazole)) as the electron acceptor. While F8BT—due to its highly luminescent character—is used in OLEDs as well, PFB is replaced by a similar donor named TFB (poly[9,9'-dioctylfluorene-co-N-(4-butylphenyl)diphenylamine]) for improved luminescence efficiency—a fact that we will shed light on later in this section.

In a first study on F8BT:PFB blends, Halls et al. studied morphological effects that arise from a complicated interplay between solidification by solvent evaporation, phase separation, and (de)wetting [235]. The authors showed that lateral phase separation can simultaneously exist on both the micrometer and nanometer scales, depending on the rate of solvent evaporation. They

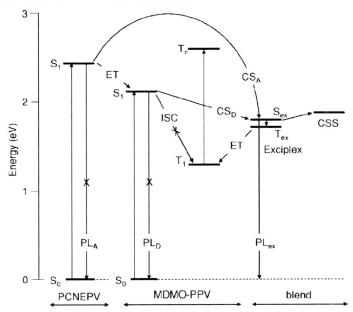

Fig. 48 Energy levels in the pristine polymers and blend system of PCNEPV and MDMO-PPV. Since the triplet state of MDMO-PPV is the lowest excited state of the system, excitations relax to it and diminish the possibility of charge carrier separation. Notation: singlet (S), triplet (T), exciplex (ex), and charge-separated states (CSS) and transitions (ET = energy transfer; CS = charge separation; PL = photoluminescence; ISC = intersystem crossing) between these states. *Crosses* indicate processes that do occur in the pure materials, but that are quenched in the blend. (Reprinted with permission from [229], © 2005, American Physical Society)

further demonstrated that the rate of solvent evaporation during spin coating can be adjusted by heating the substrate, and thereby the scale of phase separation can be controlled to some extent. As a result, the EQE could be raised by a factor of 2 from coarse grained morphologies to finer ones [235].

Arias et al. extended this study to the use of different solvents and film formation times by application of drop casting [236]. By using chloroform instead of xylene, the scale of phase separation between F8BT and PFB was restricted to the nanometer scale due to the high vapor pressure and corresponding rapid solvent evaporation of chloroform. However, a larger scale phase separation on the order of micrometers could also be obtained when the films from chloroform solution were drop cast and slowly dried in a saturated chloroform atmosphere. In general, films prepared from chloroform resulted in higher EQEs than those from xylene, with a variation corresponding to the solvent evaporation time. However, changes in the photocurrent and photoluminescence quenching efficiency were relatively small when compared to the large change in the lateral phase separation by about two orders of magnitude. Multiple scale phase separation was responsible for this: the larger, micron-sized phase domains were themselves not purely one material but exhibited some internal nanometer-sized phase separation as well (see Fig. 49). The multiple scale phase separation arose from several stages in film formation and phase separation, including heterogeneous nucleation, spinoidal decomposition, hydrodynamic regimes, coalescence, and some possible dewetting or liquid film instability effects. Finally, by combination of fluorescence microscopy, fluorescence SNOM, and tapping mode AFM the authors proved the luminescent F8BT to be the phase elevated at the surface of the film evolving in the large-scale phase separated blend films [236].

Snaith et al. investigated F8BT:PFB blends with respect to the blending ratio [237]. They found that the photoluminescence quenching efficiency was rather insensitive to the blending ratio, whereas the highest EQEs were

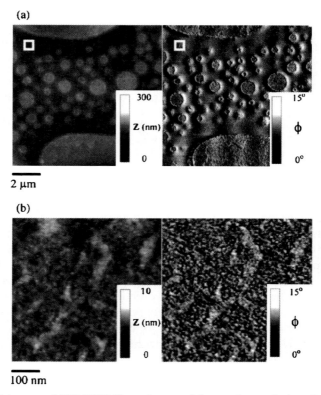

Fig. 49 AFM images of PFB:F8BT film spin-coated from xylene solution clearly demonstrating phase separation to exist on two scales. *Left*: height images; *right*: phase response images. **a** 10 μm × 10 μm and **b** 500 nm × 500 nm, which was taken in the vicinity of the sample marked by the *white square* in (**a**). (Reprinted with permission from [236], © 2004, American Chemical Society)

obtained for F8BT-enriched blends (PFB:F8BT 1 : 5). They attributed this discrepancy between the photoluminescence quenching and EQE to charge transport limitation rather than to charge generation. The authors also developed a simple model describing the interfacial area between F8BT- and PFB-rich domains and found a linear trend with respect to the EQE [237].

A further refinement of the scale of phase separation was recently demonstrated by Xia and Friend using inkjet printing (IJP) and thereby doubling the EQE [238]. As demonstrated by fluorescence and atomic force microscopies, this originates from a more rapid drying process of inkjet printed films as compared to spin cast ones (see Fig. 50). The small volume and hence the large surface to volume ratio of each IJP droplet led to this fast evaporation and drying.

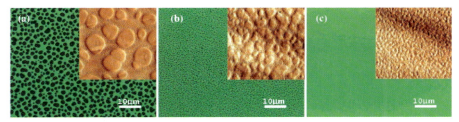

Fig. 50 Fluorescent microscopic pictures of TFB:F8BT blend films produced by spin-coating (**a**), inkjet printing at room temperature (RT) (**b**), and inkjet printing at 40 °C (**c**). The insets show the corresponding AFM pictures (10 μm × 10 μm). Blend films were deposited onto ITO-coated substrates. A reduction in scale of phase separation from **a** through **c** is demonstrated. (Reprinted with permission from [238], © 2005, American Chemical Society)

Kietzke et al. suggested another approach to control the scale of phase separation by preparation of dispersions of solid polymer nanoparticles [239, 240]. To achieve the polymer nanoparticles, either a pristine polymer or a polymer blend solution are placed into a water/surfactant mixture. Upon ultrasonication and stirring, a miniemulsion is formed that directly leads to solid polymer particles in aqueous solution by subsequent evaporation of the solvent. The route of preparation is schematically shown in Fig. 51. Indeed, the EQE spectrum of spin cast solar cells was independent of the solvent used in the miniemulsion process for blended nanoparticles [239]. Both concepts, dispersions of blends of pristine polymer particles (single PFB and single F8BT particles, separately made) as well as dispersions of blended polymer particles (PFB:F8BT blend particles), yielded comparable efficiencies to solution processed blends, with the latter yielding a threefold larger EQE [240].

In another study Arias et al. showed that a vertical phase segregation could be induced in thin films spin cast from F8BT:PFB blends by interfacial modification of the PEDOT:PSS layer with SAMs [241]. To introduce the SAM, the PEDOT:PSS layer was first modified by oxygen plasma treatment leading to

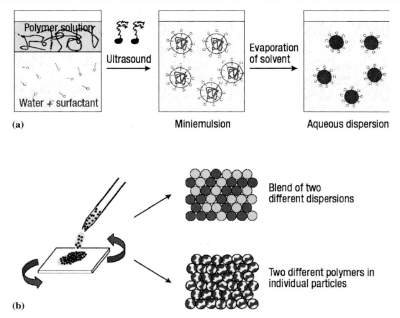

a high density of hydroxyl groups at the surface. These then acted as covalent bonding sites for 7-octenyltrichlorosilane (7-OTS), which was placed on top of the freshly treated PEDOT:PSS by an ink-stamping method (microcontact printing). With xylene as spin casting solvent, the F8BT phase did not cover the whole film surface; however, by the application of isodurene a complete coverage by an approximately 15-nm-thick F8BT layer was achieved (compare with AFM results in Fig. 52) [241].

This vertical phase segregation led to largely improved device characteristics (EQEs of 10–20%), which were accounted for by an optimized charge transport due to the F8BT being solely in contact with the electron-extracting aluminum electrode [241].

Pacios and Bradley observed bimolecular recombination for PFB:F8BT blend solar cells spin cast from chloroform, thus explaining in part the comparatively low EQEs generally observed for these devices [242].

A comprehensive discussion on the nanomorphology of polymer blends based on TFB:F8BT and PFB:F8BT, as used for polymer light-emitting diodes and polymer solar cells, respectively, is presented in [66, 67]. Kim et al. considered the ternary phase diagram, comprised of the two polymers and the

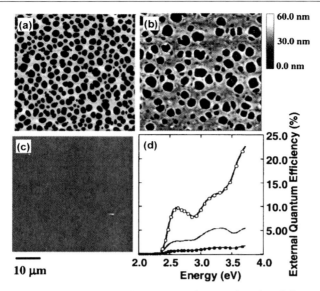

Fig. 52 Topographical AFM images of PFB:F8BT deposited under different preparation conditions onto a monolayer printed on a PEDOT electrode: **a** spin-coating from xylene solution; **b** spin-coating under a xylene-rich atmosphere from xylene solution; and **c** spin-coating from isodurene solution (leading to homogeneous F8BT coverage). **d** Short-circuit EQE action spectra of photovoltaic devices fabricated by spin-coating the blend solution from xylene (*filled circles*), under a xylene-rich atmosphere (*solid line*), and from isodurene (*open circles*). (Reprinted with permission from [241], © 2002, American Institute of Physics)

solvent, for understanding the evolving blend morphology during the spin-coating process. A schematic for the three-dimensional phase composition of the TFB:F8BT blend film during and after film formation is presented [67]. Essentially the same morphology has been found for PFB:F8BT blends, which was validated with the help of scanning Kelvin probe microscopy [243]. Thereby it was further evidenced that a PFB capping layer reduced effectively the performance of the device due to its blocking of photogenerated electrons in the F8BT-rich phase from reaching the electrode. The same effect was observed earlier in polymer–fullerene blends as well [60, 244]. Chiesa et al. further elucidated that a simple bilayer structure is advantageous for efficient electron conduction and thus device operation [243], in accordance with Arias et al. [241]. In addition Chiesa et al. detected the surface potential on either phase of the blend film to be logarithmically dependent on the impinging light intensity [243].

A logarithmic relationship between light intensity and open circuit voltage had been shown for bilayer polymer solar cells by Ramsdale et al. [125]. This and the observation of an "overpotential" of the open circuit voltage with respect to the work function difference of the two electrodes—as inferred from the MIM model [63, 245]—was explained by a charge concentration driven diffusion current in these bilayer devices (compare with Fig. 18) [125]. The observed overpotential of 1 V was practically independent of cathode work function variations through 0.8 eV. Thus, at open circuit voltage, as defined by a net zero current through the device, the field driven drift current (injection) and concentration driven diffusion current simply cancel each other. In a refined numerical model the same behavior can be closely simulated [129]. Furthermore, the authors showed that charge separation at the polymer–polymer heterojunction leads to the formation of bound polaron pairs, which may either recombine monomolecularly or be dissociated into free charges [129].

Snaith et al. related differences in open circuit voltage between bilayer and blend devices to parasitic shunt losses in the blends, when especially the PFB phase paths connect both the electrodes (see Fig. 53) [246].

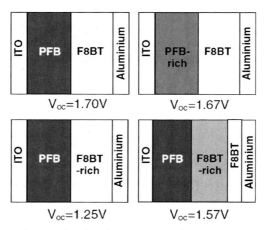

Fig. 53 Dependence of open circuit voltage V_{OC} on the architecture of PFB:F8BT solar cells. When the PFB phase can form percolation paths between both electrodes, the V_{OC} is reduced considerably. (Reproduced with permission from [246], © 2004, Wiley-VCH)

Morteani et al. demonstrated that after photoexcitation and subsequent dissociation of an exciton at the polymer–polymer heterojunction, an intermediate bound geminate polaron pair is formed across the interface [56, 57]. These geminate pairs may either dissociate into free charge carriers or collapse into an exciplex state, and either contribute to red-shifted photoluminescence or may be endothermically back-transferred to form a bulk exciton again [57]. In photovoltaic operation the first route is desired, whereas the second route is an unwanted loss channel. Figure 54 displays the potential energy curves for the different states.

The authors showed by applying the Onsager model to electric field-dependent photoluminescence quenching data [49, 247, 248] that the gemi-

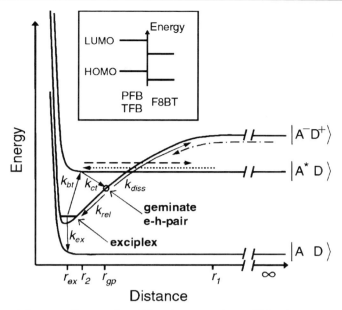

Fig. 54 Potential energy diagram describing the energetics and kinetics at type II polymer heterojunctions. The energetic order of $|A^+D^-\rangle_r = \infty$ and $|A * D\rangle_r = \infty$ may be reversed for PFB:F8BT vs TFB:F8BT. The inset shows the band offsets at a type II heterojunction. (Reprinted with permission from [57], © 2004, American Physical Society)

nate pair formed at the PFB/F8BT interface is considerably larger than that at the TFB/F8BT interface (3.1 versus 2.2 nm). Hence, TFB:F8BT blends lead to efficient light-emitting devices whereas polaron pair dissociation is largely increased for PFB:F8BT blends, as required for solar cells.

4
Organic–Inorganic Hybrid Polymer Solar Cells

Greenham et al. studied the first hybrid systems containing CdS or CdSe nanoparticles embedded in MEH-PPV [249]. As an aggregation-preventing ligand for the nanoparticles, the surfactant trioctylphosphine oxide (TOPO) was used. This surfactant, however, rather hinders charge transport between the nanoparticles and charge transfer from the conjugated polymer onto them. Further, an extension to the polymer absorption band could be achieved due to the added absorption of the nanocrystals. To reach relatively high photovoltaic performances, the system required a high load (> 80%) of nanocrystals to be incorporated, similar to the MEH-PPV:PCBM system [29]. Applying pyridine as a replacement for the TOPO coating, the first photovoltaic devices were presented (see Fig. 55).

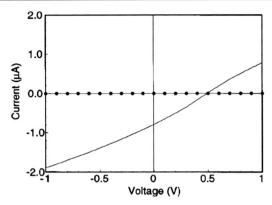

Fig. 55 Current–voltage curve for a MEH-PPV based device containing 90 wt % CdSe in the dark (*circles*) and under monochromatic illumination at 514 nm (*solid line*). The active area of the device was 7.3 mm², and the illumination was from a laser spot which was contained within the active area. The maximum power density of the illumination was approximately 5 W/m². (Reprinted with permission from [249], © 1996, American Physical Society)

A major step in the development of hybrid polymer solar cells was achieved by blending CdSe nanoparticles with regioregular P3HT. In 2002, Huynh et al. reported AM 1.5 power conversion efficiencies of 1.7% and EQEs reaching 54% with that system [250]. In this study the aspect ratio of the CdSe nanocrystals was varied roughly between 1 and 10, and the authors reported the best photovoltaic efficiencies for nanorods having the most elongated structure (7 by 60 nm) (see Fig. 56). The authors concluded that the elongated nanorod-like nanocrystals provide a better charge transport through band transport than shorter ones, where many hopping processes between the nanoparticles limit charge transport.

In a related study, Huynh et al. demonstrated control of the P3HT:CdSe nanorod blend morphology in thin films by applying a solvent mixture of chloroform and pyridine [251]. A pyridine concentration of about 8% yielded the finest intermixing of P3HT and CdSe nanorods, which was reflected by smooth film topographies. A thermal annealing step was applied at reduced pressure to remove excess pyridine from the blend, and consequently improve the EQE. The highest EQE was achieved for elongated nanorods, yielding almost 60% after thermal treatment.

Pientka et al. studied photoinduced charge transfer from CdSe and InP nanocrystals onto MDMO-PPV [252, 253]. The authors confirmed the use of pyridine to be favorable for efficient charge transfer as compared to more extended octylphosphine-containing organic ligands around the nanoparticles. These results were obtained by using photoluminescence quenching, photoinduced charge transfer, and light-induced spin resonance measurements.

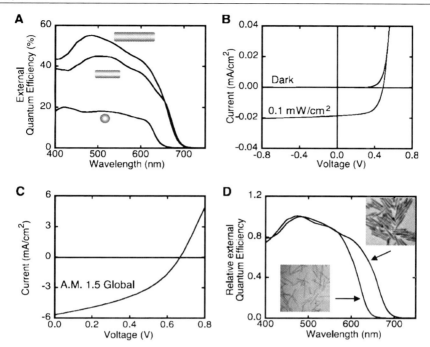

Fig. 56 ECE (**A**) of 7-, 30-, and 60-nm-long CdSe nanorods having a diameter of 7 nm. Current–voltage characteristics in the dark and 0.1 mW/cm² (**B**) and at AM 1.5 solar spectrum (**C**) of P3HT:CdSe hybrid solar cells. Photocurrent spectra of 60-nm-long nanorods with a diameter of 3 and 7 nm are compared in (**D**). From [250]. (Reprinted with permission from American Association for the Advancement of Science (AAAS), © 2002. http://www.sciencemag.org)

CdSe nanocrystal based solar cells were substantially improved by Sun et al.: a twofold increase in the EQE was achieved for MDMO-PPV based blends by application of CdSe nanotetrapods instead of nanorods [254]. The tetrapods, due to their shape, induced better directed electron transport normal to the film plane, yielding overall power conversion efficiencies of 1.8%. The current–voltage characteristics of this device are displayed in Fig. 57.

A further performance increase was achieved by application of 1,2,4-trichlorobenzene instead of chloroform as spin casting solvent: typical device efficiencies of 2.1% are reported [255]. The authors related this improvement to a vertical segregation of the CdSe tetrapods in the film. By using the same solvent for CdSe nanorods in combination with regioregular P3HT, photovoltaic devices consistently yielded power conversion efficiencies of 2.6% [36]. This has also been related to the formation of P3HT fibrils in the film, resulting in improved hole transport properties expressed by larger photocurrents and fill factors in comparison to the cases where chloroform or thiophene was used as solvent.

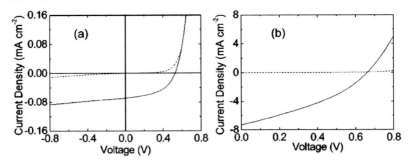

Fig. 57 a Current density vs voltage for a CdSe tetrapod/MDMO-PPV device in the dark (- - -) and under $0.39\,\text{mWcm}^{-2}$ illumination at 480 nm. $V_{OC} = 0.53\,\text{V}$, $I_{SC} = -0.069\,\text{mA cm}^{-2}$, FF = 0.49, and $\eta = 4.45\%$. **b** Current density vs voltage for the same device as in plot (**a**) illuminated with simulated AM 1.5 global light at an intensity of $93\,\text{mW cm}^{-2}$, $V_{OC} = 0.65\,\text{V}$, $I_{SC} = -7.30\,\text{mA cm}^{-2}$, FF = 0.35, and $\eta = 1.8\%$. (Reprinted with permission from [254], © 2003, American Chemical Society)

Several interesting new concepts for the design of CdSe nanocrystal based polymer solar cells have been introduced recently. Snaith et al. have infiltrated CdSe nanocrystals into polymer brushes and demonstrated EQEs of up to 50% [256]. In this case the poly(triphenylamine acrylate) (PTPAA) chains were directly grown from the substrate by a surface-initiated polymerization on tethered initiator sites (Fig. 58). The authors pronounced the wide applicability of this method for the design of nanocrystal–polymer functional blends [256].

A very controlled manner of organizing nanoparticle/copolymer mixtures was achieved by Lin et al. [257], by applying a nonconjugated block copoly-

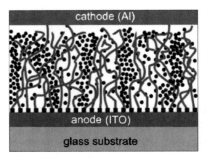

Fig. 58 Schematic of inferred structure for CdSe nanocrystal infiltrated polymer brush photovoltaic device. From *bottom* to *top*: ITO-coated glass slide modified by surface attachment of a bromine end-capped trichlorosilane self-assembled monolayer (SAM) (*squares*); polymer brushes grown from the SAM (*lines*); CdSe nanocrystals infiltrated into the brush network exhibiting some degree of phase separation in the plane of the film (*small circles*); and an aluminum cathode cap. (Reprinted with permission from [256], © 2005, American Chemical Society)

mer for a homogeneous distribution of CdSe nanoparticles in the blend film. A future application of this method to donor–acceptor block copolymers appears to be interesting.

Firth et al. have synthesized CdSe nanocrystals by application of microwave heating under mild conditions [258]. They incorporated these nanocrystals directly into blends with a copolymer based on fluorene and carbazole building blocks. Spectral photocurrent and I–V characteristics demonstrated photovoltaic operation and charge generation of both the polymer and the CdSe nanoparticles.

A unique design was proposed by Landi et al., using CdSe quantum dot–single-walled carbon nanotube complexes in blends with poly-(3-octylthiophene) (P3OT) [259]. One motivation for this construction was the ability to extend the usable absorption spectrum.

Liang et al. introduced a covalently linked layer-by-layer assembly of a PPV polymer with CdSe nanoparticles [260]. In this method the subsequent deposition of polymer and nanoparticle layers is accompanied by a covalent cross-linking at the interlayers. This resulted in a good control of total layer thickness in the device and very stable films. A first photovoltaic application was also demonstrated.

Kang et al. applied CdS nanorods in combination with MEH-PPV, demonstrating substantial improvements of these blends against MEH-PPV single layer solar cells [261]. The occurrence of photoinduced charge transfer in this system is supported by steady-state and transient photoluminescence quenching experiments. Power conversion efficiencies of 0.6% were achieved. A theoretical study on nanostructured heterojunction polymer–nanocrystal based photovoltaic devices by Kannan et al. also predicts that only small and elongated nanocrystals lead to high EQEs and thus photocurrents [262].

Many other systems based on different nanoparticles have been introduced, such as copper indium disulfide ($CuInS_2$) [263–265], copper indium diselenide ($CuInSe_2$) [266, 267], cadmium telluride (CdTe) [268], lead sulfide (PbS) [269, 270], lead selenide (PdSe) [271], and mercury telluride (HgTe) [272]. Some of these systems show enhanced spectral response well into the infrared part of the solar spectrum [271, 272]. In most cases the absorption of the nanocrystals was, however, quantitatively small as compared to the conjugated polymers.

One extensively studied material system among the nanocrystal–polymer blends is zinc oxide (ZnO) in combination with MDMO-PPV or P3HT [273–282]. Beek et al. presented the first polymer solar cells containing ZnO nanoparticles, reaching power conversion efficiencies of 1.6% [273]. In this case the nanoparticles were prepared separately and then intermixed with MDMO-PPV in solution. Shortly after this study the Janssen group presented another route to ZnO–polymer hybrid solar cells by forming the nanocrystals in situ inside the film by applying a precursor [274]. Here, diethylzinc served as the precursor and was spin cast in blends with MDMO-PPV. Process-

ing at 40% humidity and successive annealing at 110 °C yielded films with
ZnO nanoparticles approximately 6 nm in diameter intimately mixed with the
polymer. A relatively moderate volume concentration of about 15% yielded
the best device results, reaching 1.1% power conversion efficiency [274]. In
the case of separately prepared and intermixed ZnO nanoparticles, it was
shown that about 30% by volume of ZnO is required for optimization of
device parameters [275]. Application of ZnO nanorods in the same devices
yielded slight improvements in fill factor [275]. In combination with P3HT
the same ZnO nanoparticles, however, showed a lower performance than
that for MDMO-PPV [277]. This has been attributed to the coarseness of the
morphology in the blend films.

Another interesting concept is the application of ZnO nanofibers grown
from the substrate [280–282]. Best power conversion efficiencies of 0.5% were
limited in their photocurrent by a rather large spacing between the separate ZnO nanofibers (Fig. 59) [280]. Furthermore, it was shown that the ZnO

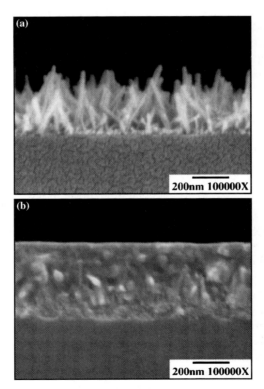

Fig. 59 a SEM image of a glass/ZnO nucleation layer/ZnO nanocarpet structure. The ZnO
nanofibers are grown from an aqueous solution of zinc nitrate, and the nucleation layer
is spin-coated from a zinc acetate solution. **b** SEM image of P3HT intercalated into the
nanocarpet structure. (Reprinted from [280], © 2006, with permission from Elsevier)

nanofiber growth can be controlled by a seed layer, and the application of an amphiphilic molecule as interface layer can help to reduce charge recombination [281, 282].

Closely related to liquid electrolyte dye-sensitized solar cells (DSSCs, also known as "Grätzel cells") [283, 284], the class of solid-state DSSCs has been developed to improve device stability and reduce complications in the production process [285–288]. Thus, although polymers can be utilized as replacements for sensitizing dyes (as in liquid electrolyte DSSCs) [289–291], the main effort in applying conjugated polymers focuses on solid-state DSSCs [45, 292–298]. With environmentally friendly production of this polymer based solid-state DSSC in mind, a device based on water-soluble polythiophene derivative has been presented as well [299].

Rather sophisticated structures of the TiO_2 porous film were introduced by Coakley et al. [69, 70]. Based on a titanium(IV) tetraethoxide (TEOT) titania precursor in combination with a pluronic poly(ethylene oxide)–poly(propylene oxide)–poly(ethylene oxide) triblock copolymer (P123) as the structure-directing agent, regular hexagonal structures (honeycomb) with pore sizes around 10 nm were prepared (Fig. 60).

Fig. 60 High-resolution SEM top view image of a mesoporous TiO_2 film following calcination at 400 °C. The pore diameter in the plane of the film is ~10 nm. (Reprinted with permission from [70], © 2003, American Institute of Physics)

After production of the mesoporous TiO_2 film, regioregular P3HT was infiltrated into these pores followed by an annealing step. Even for the highest annealing temperatures of 200 °C the authors estimated a total polymer infiltration of only 33%, since in the small pores the polymer has to form rather coiled structures as indicated by a blue-shift in the polymer absorption and emission spectra. This coiling was proposed to be a reason for lower exciton and charge transport efficiencies, leading to estimated power conversion efficiencies somewhat below 0.5% [70]. For larger pore diameters up to 80 nm obtained in anodic alumina, Coakley et al. observed an increase of the P3HT hole mobility due to alignment of the polymer chains along the pores [300].

Bartholomew and Heeger reported recently on the infiltration of P3HT in random nanocrystalline TiO$_2$ networks [301]. The TiO$_2$ networks were produced by spin-coating TiO$_2$ nanocrystals modified by organics from dispersion. Due to a low resulting porosity of the TiO$_2$ film (Fig. 61), infiltration of the polymer appeared to be difficult. Yet, the amount of infiltrated P3HT could be effectively increased by using a lower molecular weight fraction of the polymer, in combination with annealing and surface modification of the TiO$_2$ nanocrystals by applying amphiphilic Ru-based dyes [301].

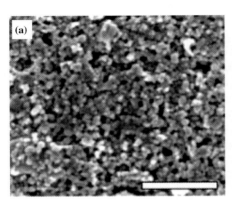 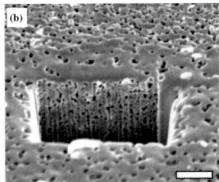

Fig. 61 SEM image of a random nanocrystalline TiO$_2$ film created by spin-coating. **a** Top view and **b** angled view of a cross section. The scale bars represent 500 nm. (Reproduced with permission from [301], © 2005, Wiley-VCH)

Solid-state DSSCs have much lower efficiencies as compared to liquid electrolyte Grätzel cells, most probably due to charge transport and recombination limitations. Ravirajan et al. reported that an additional PEDOT:PSS layer under the hole-extracting gold electrode improved charge extraction, leading to overall power conversion efficiencies of about 0.6% [302]. Further increase in the power conversion efficiency (0.7%) was reached for similarly constructed devices, where the porous TiO$_2$ layer was optimized in its interconnecting network structure by applying structure-directing polystyrene-*block*-polyethylene oxide diblock copolymer templates (P(S-*b*-EO); Fig. 62) [303].

A straightforward approach to forming a conjugated polymer/nanocrystalline TiO$_2$ hybrid bulk heterojunction was reported by van Hal et al. [304]. The TiO$_2$ precursor titanium(IV) isopropoxide (Ti(OC$_3$H$_7$)$_4$) was spin-coated in a THF solution directly together with the conjugated polymer MDMO-PPV to form intermixed thin films. Subsequently the TiO$_2$ precursor was converted into nanocrystalline TiO$_2$ via hydrolysis in air. This resulted in a hybrid bulk heterojunction with characteristic domain sizes of about 50 nm. Slooff et al. showed that the relative humidity in air has a major influence on the morphology formed within bare TiO$_2$ and blend films prepared by hydrolysis (Fig. 63) [305]. Furthermore, the authors identified the low crystallinity

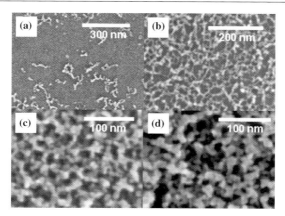

Fig. 62 SEM images of P(S-*b*-EO) and titanium tetraisopropoxide (TTIP) films with TTIP in different solvents annealed at 400 °C for 5 h: **a** isopropanol, **b** xylene, **c** chloroform, and **d** chlorobenzene. (Reprinted with permission from [303], © 2006, Institute of Physics Publishing)

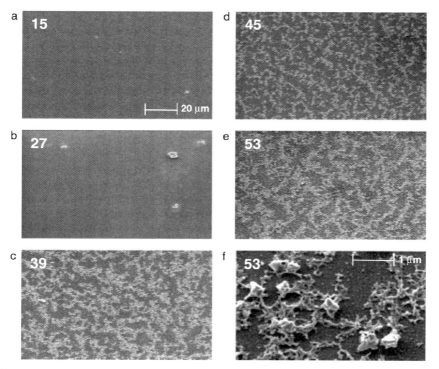

Fig. 63 Low-voltage scanning electron microscopy of TiO_2 films spin-coated at different relative humidities (RH). Images **a–e** have identical scale, while **f** shows a higher magnification of the film spin-coated at 53% RH. (Reproduced with permission from [305], © 2005, Wiley VCH)

of the TiO$_2$ in the blends as a bottleneck for electron transport through the device.

Another interesting approach to form P3HT/TiO$_2$ bulk heterojunctions was proposed by Feng et al. [306]. They used ultrasonic-assisted polymerization of 3-hexylthiophene directly into dispersed TiO$_2$ nanocrystals in chloroform, followed by spin casting of thin films. The authors achieved more homogeneous films by increasing the weight fraction of TiO$_2$ to an optimal ratio of 50%. Absorption spectra indicate a rather ordered polymer structure on top of the nanocrystals leading to improved charge generation.

5
Carbon Nanotubes in Polymer Solar Cells

Since their discovery in 1991, carbon nanotubes (CNTs) have been a constant source of scientific inspiration [307]. Among the most intriguing properties of CNTs is the electric field enhanced electron emission from the nanotubes [308]. Initial studies combining CNTs and conjugated polymers concentrated on the diode properties in a CNT–polymer heterojunction [309]. Romero et al. demonstrated light-sensitive photodetectors in combination with a PPV derivative. The authors showed that hole injection from the CNT electrode proved to be much more efficient than using an ITO electrode, and related this phenomenon to the enhancement of the local electric field at the tip of the nanotubes [309].

Curran et al. demonstrated the use of multiwalled carbon nanotubes (MWNTs) to increase the conductivity within PPV films. Blending about 15% (by mass) CNTs into the PPV films yielded improvements of five orders of magnitude and was accompanied by a reduction in photoluminescence efficiency [310].

The application of MWNTs in the field of polymer solar cells was presented by Ago et al. [311]. Here, a layer of CNTs served as a replacement for the common ITO hole-collecting electrode in a single layer PPV/Al diode (Fig. 64). The authors related the twofold enhancement of the EQE observed with the MWNT based device to the formation of a complex network with an increased interface area between MWNTs and PPV, in addition to a stronger built-in electric field as a result of the higher work function of MWNTs compared to the standard ITO electrode [311, 312]. To determine whether electron or energy transfer processes dominate within MWNT:PPV blends, their photophysical properties were studied by photoluminescence and PIA spectroscopy. The results confirmed nonradiative energy transfer from PPV singlet excitons to the MWNTs as the main electronic interaction [313].

The first single-walled carbon nanotube (SWNT)–conjugated polymer photovoltaic devices were presented by Kymakis et al. [314]. As photoactive

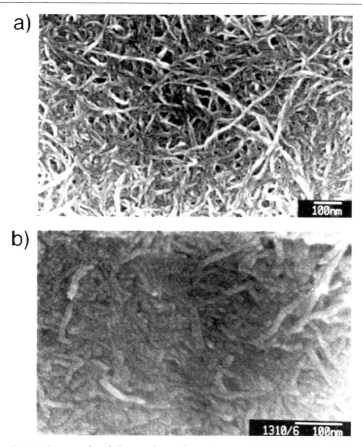

Fig. 64 a SEM micrograph of the surface of a spin-coated MWNT film. **b** The cleaved surface of a PPV±MWNT composite near the bottom of the MWNT layer. The small particles seen in the composite are evaporated gold particles used to avoid a charging effect because the composite is less conducting than the pure MWNT film. (Reproduced with permission from [311], © 1999, Wiley VCH)

layer a blend of P3OT with a low SWNT concentration (< 1% by weight) was sandwiched between an ITO and an aluminum electrode. Thereby the CNT–P3OT junctions acted as dissociation centers for excitons on the polymer, enabling efficient transport of electrons via the nanotubes to the metal electrode [314]. The power conversion efficiency increased in comparison to the pristine polymer film by about three orders of magnitude. The observation of a relatively high open circuit voltage (V_{OC}) of 0.75 V was attributed to the formation of ohmic contacts between the CNTs and the metal electrode, resulting in a weak dependence of V_{OC} on the metal electrode used [315]. While dye functionalization of CNTs yielded a relative increase of the observed photocurrent [316], the application of PEDOT:PSS modified ITO electrodes yielded power conversion efficiencies of 0.1% [317]. Further, it was shown that around only 1% mass fraction of SWNTs in P3OT yielded the highest photocurrents [317]. To date, record efficiencies of 0.22% were recently achieved by application of a thermal annealing step to the identical system (SWNT/P3OT) (Fig. 65) [37].

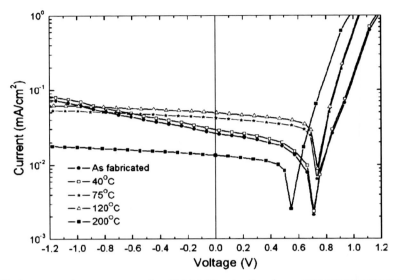

Fig. 65 Current–voltage curves under AM 1.5 illumination for an ITO/PEDOT:PSS/P3OT-SWNT cell with different postfabrication annealing temperatures. Optimal performance is achieved for annealing temperatures of 120 °C. (Reprinted with permission from [37], © 2006, Institute of Physics Publishing)

Photocurrent response in the near-infrared region up to 1600 nm, related to absorption features of semiconducting SWNTs in blends with MEH-PPV and P3OT, offers principally the operation of infrared sensitive photodetectors with these materials [318]. To enable the photon harvesting in this spectral region, the SWNTs needed to be finely dispersed within the polymer matrices, thereby switching off the excitation quenching observed within CNT bundles.

Applying a layer-by-layer (LBL) deposition technique of polythiophenes bearing carboxylic groups on alkylsulfanyl side chains in combination with pyrene⁺-modified SWNTs, Rahman et al. reported monochromatic power conversion efficiencies of more than 9% for eight subsequently deposited sandwich layers [319]. Guldi et al. suggested supramolecular structures employing an electrostatic interaction between pyrene or PSS substituted SWNTs and charged porphyrins as building blocks for solar energy conversion [320].

Landi et al. applied laser pulse vaporized production of SWNTs for use in blends with P3OT [321]. Upon formation of a two-layer system consisting of a pristine P3OT film on ITO followed by deposition of a SWNT:P3OT blend film, the authors observed an unusually high open circuit voltage of about 1 V (under AM 0 illumination) (Fig. 66).

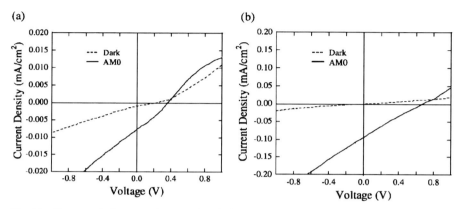

Fig. 66 Characteristic $I-V$ plots in the dark (*dotted line*) and under simulated AM 0 illumination (*full line*) displaying the photoresponse for: **a** pristine P3OT; **b** 1% w/w SWNT–P3OT composite solar cells. (Reproduced with permission from [321], © 2005, John Wiley & Sons, Ltd.)

Although Itoh et al. did not observe an improvement of photovoltaic devices based on a bilayer configuration of TiO_2 and P3OT via doping of the P3OT layer with SWNTs, they did find a dramatic increase of the current under reverse bias. This was attributed to field emission of electrons from the CNT onto the dense TiO_2 film [322].

Rud et al. applied electric fields for the improved vertical orientation of SWNTs in a blend with a water-soluble polythiophene derivative [323]. A profound increase of both the conductance and the photocurrent was observed for the devices where CNTs were aligned. The orientation of MWNTs in composites with polymers can also be influenced by application of large magnetic fields [324].

CNT–CdS complexes have been suggested for solar energy conversion application [325, 326], and similar complexes based on CdSe were applied as well [259].

Pradhan et al. functionalized MWNTs with ester groups in order to better disperse them into P3HT in a bilayer device with C_{60} on top [327]. Upon comparison with undoped P3HT layers, the authors identified a threefold effect of the CNTs: (a) increase of the open circuit voltage V_{OC} due to the work function of the CNT, (b) sites for P3HT exciton dissociation, and (c) efficient pathways for hole transport (besides the P3HT) to the ITO electrode [327].

A remarkably large open circuit voltage was recently observed by Patyk et al. for P3OT:SWNT blends prepared on top of an electrochemically deposited polybithiophene layer [328]. Beforehand the SWNTs were modified by 2-(2-thienyl)ethanol groups. Use of Ca/Al metal back electrodes resulted in a remarkable enhancement of the V_{OC} of up to 1.8 V as compared to the bare aluminum contact showing 1 V.

Deeper insight into CNT functionalities when used as a dopant for polymer solar cells and polymer light-emitting diodes (PLEDs) was presented by Xu et al. [329]. While the PLED gained from rather low CNT doping levels of about 0.02%, the solar cell performance increased further up to 0.2 wt %. The improved EQE of the OLED was explained by a better charge carrier injection from the electrode, whereas for the solar cell exciton dissociation is facilitated by the nanotubes [329].

Recently the idea of Ago et al. to replace the ITO electrode by a CNT based electrode was pursued by several groups again. However, this time SWNTs were used [330–333]. The motivation for this step is generally found in the benefit of replacing an expensive vacuum step in the fabrication of polymer solar cells [330] with roll-to-roll production of supporting nanotube electrodes (Fig. 67) [331], which will aid in the removal of ITO and PEDOT:PSS related problems [332] while facilitating applications of flexible devices on plastic substrates [333].

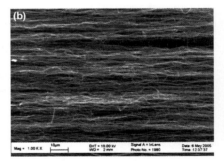

Fig. 67 a SEM of the CNT sheet being dry drawn from a CNT forest into a self-assembled sheet. **b** Undensified, dry-drawn single layer sheet of free standing CNTs. (Reproduced with permission from [331], © 2006, Wiley-VCH)

Rowell et al. demonstrated that SWNT based hole-collecting electrodes show well above 80% of the performance of classical ITO electrodes, and they allow a significantly higher bending stress (hence, greater flexibility) than ITO based plastic substrates [333]. Thus, corrected power conversion efficiencies of 2.5% were presented and the slight power loss in comparison to test devices using ITO contacts was mainly attributed to the increased serial resistance [333]. The current–voltage characteristics of devices with SWNT or ITO electrodes are compared in Fig. 68.

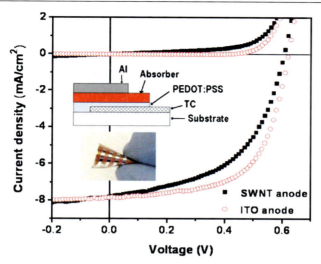

Fig. 68 Current density–voltage characteristics of P3HT:PCBM devices under AM 1.5G conditions using ITO on glass (*open circles*) and flexible SWNTs on PET (*solid squares*) as the anodes, respectively. Insets: schematic of device and photograph of the highly flexible cell using SWNTs on PET. (Reprinted with permission from [333], © 2006, American Institute of Physics)

6
Conclusions and Outlook

Several device concepts employing conjugated polymers as active components in the photoconversion process of photovoltaic devices have been presented to date. With power conversion efficiencies surpassing 5% (polymer–fullerene), reaching 3% (hybrid polymer–nanoparticle), or 2% (polymer–polymer), the prospects are high.

Intensified synthetic efforts are needed, especially for lowering the absorption band edge (low band gap) for increased photocurrent generation and simultaneously keeping the polymer HOMO level well below 5 eV for high open circuit voltages. Further control of the favorably crystalline order in the polymer will be another aspect to be envisioned for pushing power conversion efficiencies toward 10%.

The ideal schematic structure of a bulk heterojunction solar cell is displayed in Fig. 69. The donor and acceptor phases are interspaced by around 10–20 nm—comparable to the exciton diffusion length. The interdigitated and percolated "highways" ensure unhindered charge carrier transport. Last but not least, a pure donor phase at the hole-collecting electrode and a pure acceptor phase at the electron-collecting electrode have to be placed, thereby minimizing losses of recombination of the wrong sign of charges at the wrong electrode. Such a well-organized nanostructure is not easy to obtain due to

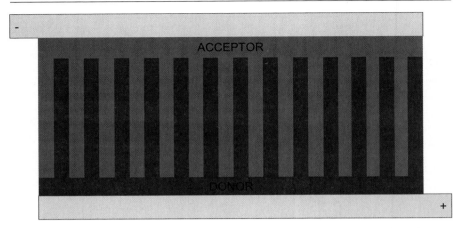

Fig. 69 Ideal structure of a donor–acceptor bulk heterojunction polymer solar cell

disorder. However, self-organization of the organic semiconducting polymers (molecules) is a key to nanoscale order.

In conclusion, this field of polymer solar cells requires high interdisciplinarity between macromolecular chemistry, supramolecular chemistry, physical chemistry, colloid chemistry, photophysics/photochemistry, device physics, nanostructural analysis, and thin film technology.

Acknowledgements HH would like to thank A.J. Ledbetter for useful discussions.

References

1. Nalwa HS (ed) (1997) Handbook of organic conductive molecules and polymers, vols 1–4. Wiley, Chichester
2. Hadziioannou G, van Hutten PF (eds) (2000) Semiconducting polymers, vol 1. Wiley-VCH, Weinheim
3. Skotheim TA, Reynolds JR (eds) (2006) Handbook of conducting polymers, vols 1–2. CRC, Boca Raton
4. McGehee MD, Miller EK, Moses D, Heeger AJ (1999) In: Bernier P, Lefrant S, Bidan G (eds) Advances in synthetic metals: twenty years of progress in science and technology. Elsevier, Lausanne, p 98
5. Dimitrakopoulos CD, Mascaro DJ (2001) Organic thin-film transistors: a review of recent advances. IBM J Res Dev 45:11
6. Hoppe H, Arnold N, Meissner D, Sariciftci NS (2003) Modeling the optical absorption within conjugated polymer/fullerene-based bulk-heterojunction organic solar cells. Sol Energy Mater Sol Cells 80:105
7. Hoppe H, Sariciftci NS (2004) Organic solar cells: an overview. J Mater Res 19:1924
8. Montanari I, Nogueira AF, Nelson J, Durrant JR, Winder C, Loi MA, Sariciftci NS, Brabec CJ (2001) Transient optical studies of charge recombination dynamics in a polymer/fullerene composite at room temperature. Appl Phys Lett 81:3001

9. Nogueira AF, Montari I, Nelson J, Durrant JR, Winder C, Sariciftci NS, Brabec C (2003) Charge recombination in conjugated polymer/fullerene blended films studied by transient absorption spectroscopy. J Phys Chem B 107:1567

10. Nelson J (2003) Diffusion-limited recombination in polymer–fullerene blends and its influence on photocurrent collection. Phys Rev B 67:155209

11. Sirringhaus H, Brown PJ, Friend RH, Nielsen MM, Bechgaard K, Langeveld-Voss BMW, Spiering AJH, Janssen RAJ, Meijer EW, Herwig P, de Leeuw DM (1999) Two-dimensional charge transport in self-organized, high-mobility conjugated polymers. Nature 401:685

12. Singh Th B, Marjanovic N, Matt GJ, Günes S, Sariciftci NS, Montaigne Ramil A, Andreev A, Sitter H, Schwödiauer R, Bauer S (2004) High-mobility n-channel organic field-effect transistors based on epitaxially grown C_{60} films. Org Electron 6:105

13. Brabec CJ, Sariciftci NS, Hummelen JC (2001) Plastic solar cells. Adv Funct Mater 11:15

14. Brabec CJ, Dyakonov V, Parisi J, Sariciftci NS (eds) (2003) Organic photovoltaics: concepts and realization, vol 60. Springer, Berlin

15. Brabec CJ (2004) Organic photovoltaics: technology and market. Sol Energy Mater Sol Cells 83:273

16. Mozer AJ, Sariciftci NS (2006) In: Skotheim TA, Reynolds JR (eds) Conjugated polymers: processing and applications, vol 2. CRC, Boca Raton, p 101

17. Science Citation Index, polymer solar cell(s) (2007) Thompson Scientific, Web of Science

18. Ma W, Yang C, Gong X, Lee K, Heeger AJ (2005) Thermally stable, efficient polymer solar cells with nanoscale control of the interpenetrating network morphology. Adv Funct Mater 15:1617

19. Reyes-Reyes M, Kim K, Carroll DL (2005) High-efficiency photovoltaic devices based on annealed poly(3-hexylthiophene) and 1-(3-methoxycarbonyl)-propyl-1-phenyl-(6,6)C61 blends. Appl Phys Lett 87:083506

20. Reyes-Reyes M, Kim K, Dewald J, Lopez R-S, Avadhanula A, Curran S, Carroll DL (2005) Meso-structure formation for enhanced organic photovoltaic cells. Org Lett 7:5749

21. Li G, Shrotriya V, Huang J, Yao Y, Moriarty T, Emery K, Yang Y (2005) High-efficiency solution processable polymer photovoltaic cells by self-organization of polymer blends. Nat Mater 4:864

22. Karg S, Riess W, Dyakonov V, Schwoerer M (1993) Electrical and optical characterization of poly(phenylene-vinylene) light emitting diodes. Synth Met 54:427

23. Yu G, Zhang C, Heeger AJ (1994) Dual-function semiconducting polymer devices: light-emitting and photodetecting diodes. Appl Phys Lett 64:1540

24. Yu G, Pakbaz K, Heeger AJ (1994) Semiconducting polymer diodes: large size, low cost photodetectors with excellent visible–ultraviolet sensitivity. Appl Phys Lett 64:3422

25. Marks RN, Halls JJM, Bradley DDC, Friend RH, Holmes AB (1994) The photovoltaic response in poly(p-phenylene vinylene) thin-film devices. J Phys Condens Matter 6:1379

26. Sariciftci NS, Smilowitz L, Heeger AJ, Wudl F (1992) Photoinduced electron transfer from a conducting polymer to buckminsterfullerene. Science 258:1474

27. Sariciftci NS, Braun D, Zhang C, Srdanov VI, Heeger AJ, Stucky G, Wudl F (1993) Semiconducting polymer–buckminsterfullerene heterojunctions: diodes, photodiodes, and photovoltaic cells. Appl Phys Lett 62:585

28. Yu G, Heeger AJ (1995) Charge separation and photovoltaic conversion in polymer composites with internal donor/acceptor heterojunctions. J Appl Phys 78:4510

29. Yu G, Gao J, Hummelen JC, Wudl F, Heeger AJ (1995) Polymer photovoltaic cells: enhanced efficiencies via a network of internal donor–acceptor heterojunctions. Science 270:1789
30. Halls JJM, Walsh CA, Greenham NC, Marseglia EA, Friend RH, Moratti SC, Holmes AB (1995) Efficient photodiodes from interpenetrating polymer networks. Nature 376:498
31. Tada K, Hosada K, Hirohata M, Hidayat R, Kawai T, Onoda M, Teraguchi M, Masuda T, Zakhidov AA, Yoshino K (1997) Donor polymer (PAT6)–acceptor polymer (CNPPV) fractal network photocells. Synth Met 85:1305
32. Granström M, Petritsch K, Arias AC, Lux A, Andersson MR, Friend RH (1998) Laminated fabrication of polymeric photovoltaic diodes. Nature 395:257
33. Jenekhe SA, Yi S (2000) Efficient photovoltaic cells from semiconducting polymer heterojunctions. Appl Phys Lett 77:2635
34. Shaheen SE, Brabec CJ, Sariciftci NS, Padinger F, Fromherz T, Hummelen JC (2001) 25% efficient organic plastic solar cells. Appl Phys Lett 78:841
35. McNeill CR, Abrusci A, Zaumseil J, Wilson R, McKiernan MJ, Burroughes JH, Halls JJM, Greenham NC, Friend RH (2007) Dual electron donor/electron acceptor character of a conjugated polymer in efficient photovoltaic diodes. Appl Phys Lett 90:193506
36. Sun B, Greenham NC (2006) Improved efficiency of photovoltaics based on CdSe nanorods and poly(3-hexylthiophene) nanofibers. Phys Chem Chem Phys 8:3557
37. Kymakis E, Koudoumas E, Franghiadakis I, Amaratunga GAJ (2006) Post-fabrication annealing effects in polymer–nanotube photovoltaic cells. J Phys D Appl Phys 39:1058
38. de Boer B, Stalmach U, van Hutten PF, Melzer C, Krasnikov VV, Hadziioannou G (2001) Supramolecular self-assembly and opto-electronic properties of semiconducting block copolymers. Polymer 42:9097
39. Gratt JA, Cohen RE (2004) The role of ordered block copolymer morphology in the performance of organic/inorganic photovoltaic devices. J Appl Polym Sci 91:3362–3368
40. Sun S, Fan Z, Wang Y, Haliburton J (2005) Organic solar cell optimizations. J Mater Sci 40:1429
41. Heiser T, Adamopoulos G, Brinkmann M, Giovanella U, Ould-Saad S, Brochon C, van de Wetering K, Hadziioannou G (2006) Nanostructure of self-assembled rod–coil block copolymer films for photovoltaic application. Thin Solid Films 511–512:219
42. Lindner SM, Hüttner S, Chiche A, Thelakkat M, Krausch G (2006) Charge separation at self-assembled nanostructured bulk interface in block copolymers. Angew Chem Int Ed 45:3364–3368
43. Sommer M, Lindner SM, Thelakkat M (2007) Microphase-Separated Donor–Acceptor Diblock Copolymers: Influence of HOMO Energy Levels and Morphology on Polymer Solar Cells. Adv Funct Mater 17:1493–1500
44. Halls JJM, Pichler K, Friend RH, Moratti SC, Holmes AB (1996) Exciton diffusion and dissociation in a poly(p-phenylenevinylene)/C_{60} heterojunction photovoltaic cell. Appl Phys Lett 68:3120
45. Savanije TJ, Warman JM, Goossens A (1998) Visible light sensitisation of titanium dioxide using a phenylene vinylene polymer. Chem Phys Lett 287:148
46. Pettersson LAA, Roman LS, Inganäs O (1999) Modeling photocurrent action spectra of photovoltaic devices based on organic thin films. J Appl Phys 86:487
47. Haugeneder A, Neges M, Kallinger C, Spirkl W, Lemmer U, Feldmann J, Scherf U, Harth E, Gügel A, Müllen K (1999) Exciton diffusion and dissociation in conjugated polymer/fullerene blends and heterostructures. Phys Rev B 59:15346

48. Stoessel M, Wittmann G, Staudigel J, Steuber F, Blässing J, Roth W, Klausmann H, Rogler W, Simmerer J, Winnacker A, Inbasekaran M, Woo EP (2000) Cathode-induced luminescence quenching in polyfluorenes. J Appl Phys 87:4467
49. Pope M, Swenberg CE (1999) Electronic processes in organic crystals and polymers, 2nd edn. Oxford University Press, New York
50. Sariciftci NS (ed) (1997) Primary photoexcitations in conjugated polymers: molecular exciton versus semiconductor band model. World Scientific, Singapore
51. Chandross M, Mazumdar S, Jeglinski S, Wei X, Vardeny ZV, Kwock EW, Miller TM (1994) Excitons in poly(*para*-phenylenevinylene). Phys Rev B 50:14702
52. Campbell IH, Hagler TW, Smith DL, Ferraris JP (1996) Direct measurement of conjugated polymer electronic excitation energies using metal/polymer/metal structures. Phys Rev Lett 76:1900
53. Knupfer M (2003) Exciton binding energies in organic semiconductors. Appl Phys A 77:623
54. Brabec CJ, Zerza G, Cerullo G, Silvestri SD, Luzzati S, Hummelen JC, Sariciftci S (2001) Tracing photoinduced electron transfer process in conjugated polymer/fullerene bulk heterojunctions in real time. Chem Phys Lett 340:232
55. Hoppe H, Niggemann M, Winder C, Kraut J, Hiesgen R, Hinsch A, Meissner D, Sariciftci NS (2004) Nanoscale morphology of conjugated polymer/fullerene based bulk-heterojunction solar cells. Adv Funct Mater 14:1005
56. Morteani AC, Dhoot AS, Kim JS, Silva C, Greenham NC, Friend RH, Murphy C, Moons E, Ciná S, Burroughes JH (2003) Barrier-free electron–hole capture in polymer blend heterojunction light-emitting diodes. Adv Mater 15:1708
57. Morteani AC, Sreearunothai P, Herz LM, Phillips RT, Friend RH, Silva C (2004) Exciton regeneration at polymeric semiconductor heterojunctions. Phys Rev Lett 92:247240
58. Mihailetchi VD, Koster LJA, Hummelen JC, Blom PWM (2004) Photocurrent generation in polymer–fullerene bulk heterojunctions. Phys Rev Lett 93:216601
59. Koster LJA, Smiths ECP, Mihailetchi VD, Blom PWM (2005) Device model for the operation of polymer/fullerene bulk heterojunction solar cells. Phys Rev B 72:085205
60. Hoppe H, Glatzel T, Niggemann M, Hinsch A, Lux-Steiner MC, Sariciftci NS (2005) Kelvin probe force microscopy study on conjugated polymer/fullerene bulk heterojunction organic solar cells. Nano Lett 5:269
61. Hoppe H, Glatzel T, Niggemann M, Schwinger W, Schaeffler F, Hinsch A, Lux-Steiner MC, Sariciftci NS (2006) Efficiency limiting morphological factors of MDMO-PPV:PCBM plastic solar cells. Thin Solid Films 511–512:587
62. Hoppe H, Sariciftci NS (2006) Morphology of polymer/fullerene bulk heterojunction solar cells. J Mater Chem 16:45
63. Sze SM (1981) Physics of semiconductor devices. Wiley, New York
64. Parker ID (1994) Carrier tunneling and device characteristics in polymer light-emitting diodes. J Appl Phys 75:1656
65. Tang CW (1986) Two-layer organic photovoltaic cell. Appl Phys Lett 48:183
66. Moons E (2002) Conjugated polymer blends: linking film morphology to performance of light emitting diodes and photodiodes. J Phys Condens Matter 14:12235–12260
67. Kim J-S, Ho PKH, Murphy CE, Friend RH (2004) Phase separation in polyfluorene-based conjugated polymer blends: lateral and vertical analysis of blend spin-cast thin films. Macromolecules 37:2861
68. Yang X, van Duren JKJ, Janssen RAJ, Michels MAJ, Loos J (2004) Morphology and thermal stability of the active layer in poly(*p*-phenylenevinylene)/methanofullerene plastic photovoltaic devices. Macromolecules 37:2151

69. Coakley KM, Liu Y, McGehee MD, Frindell K, Stucky GD (2003) Infiltrating semi-conducting polymers into self-assembled mesoporous titania films for photovoltaic applications. Adv Funct Mater 13:301

70. Coakley KM, McGehee MD (2003) Photovoltaic cells made from conjugated polymers infiltrated into mesoporous titania. Appl Phys Lett 83:3380

71. Coakley KM, McGehee MD (2004) Conjugated polymer photovoltaic cells. Chem Mater 16:4533

72. Campbell IH, Rubin S, Zawodzinski TA, Kress JD, Martin RL, Smith DL, Barashkov NN, Ferraris JP (1996) Controlling Schottky energy barriers in organic electronic devices using self-assembled monolayers. Phys Rev B 54:14321

73. Campbell IH, Kress JD, Martin RL, Smith DL, Barashkov NN, Ferraris JP (1997) Controlling charge injection in organic electronic devices using self-assembled monolayers. Appl Phys Lett 71:3528

74. Ganzorig C, Matsuda Y, Fujihira M (2002) Chemical modification of indium-tin-oxide electrodes by surface molecular design. Presented at the Materials Research Society meeting, Warrendale. MRS Proc 708:BB3221

75. Lee J, Jung B-J, Lee J-I, Chu HY, Do L-M, Shim H-K (2002) Modification of an ITO anode with a hole-transporting SAM for improved OLED device characteristics. J Mater Chem 12:3494

76. Scott JC, Carter SA, Karg S, Angelopoulos M (1997) Polymeric anodes for organic light-emitting diodes. Synth Met 85:1197

77. Cao Y, Yu G, Zhang C, Menon R, Heeger AJ (1997) Polymer light-emitting diodes with polyethylene dioxythiophene–polystyrene sulfonate as the transparent anode. Synth Met 87:171

78. Kugler T, Salaneck WR, Rost H, Holmes AB (1999) Polymer band alignment at the interface with indium tin oxide: consequences for light emitting devices. Chem Phys Lett 310:391

79. Greczynski G, Kugler T, Salaneck WR (2000) Energy level alignment in organic-based three-layer structures studied by photoelectron spectroscopy. J Appl Phys 88:7187

80. Hung LS, Tang CW, Mason MG (1997) Enhanced electron injection in organic electroluminescence devices using an Al/LiF electrode. Appl Phys Lett 70:152

81. Jabbour GE, Kawabe Y, Shaheen SE, Wang JF, Morrell MM, Kippelen B, Peyghambarian N (1997) Highly efficient and bright organic electroluminescent devices with an aluminum cathode. Appl Phys Lett 71:1762

82. Shaheen SE, Jabbour GE, Morrell MM, Kawabe Y, Kippelen B, Peyghambarian N, Nabor M-F, Schlaf R, Mash EA, Armstrong NR (1998) Bright blue organic light-emitting diode with improved color purity using a LiF/Al cathode. J Appl Phys 84:2324

83. Jabbour GE, Kippelen B, Armstrong NR, Peyghambarian N (1998) Aluminum based cathode structure of enhanced electron injection in electroluminescent organic devices. Appl Phys Lett 73:1185

84. Koch N, Pogantsch A, List EJW, Leising G, Blyth RIR, Ramsey MG, Netzer FP (1999) Low-onset organic blue light emitting devices obtained by better interface control. Appl Phys Lett 74:2909

85. Cao Y, Yu G, Parker ID, Heeger AJ (2000) Ultrathin layer alkaline earth metals as stable electron-injection electrodes for polymer light emitting diodes. J Appl Phys 88:3618

86. Heil H, Steiger J, Karg S, Gastel M, Ortner H, von Seggern H, Stößel M (2001) Mechanisms of injection enhancements in organic light-emitting diodes through an Al/LiF electrode. J Appl Phys 89:420

87. Li F, Feng J, Cheng G, Liu S (2002) Electron injection and electroluminescence investigation of organic light-emitting devices based on a Sn/Al cathode. Synth Met 126:347

88. Malliaras GG, Scott JC (1998) The roles of injection and mobility in organic light emitting diodes. J Appl Phys 83:5399

89. Roman LS, Mammo W, Petterson LAA, Andersson MR, Inganäs O (1998) High quantum efficiency polythiophene/C_{60} photodiodes. Adv Mater 10:774

90. Shaheen SE, Brabec CJ, Sariciftci NS, Jabbour GE (2001) Effects of inserting highly polar salts between the cathode and active layer of bulk heterojunction photovoltaic devices. Presented at the Materials Research Society meeting, San Francisco. MRS Proc 665:C5511

91. Brabec CJ, Shaheen SE, Winder C, Sariciftci NS, Denk P (2002) Effect of LiF/metal electrodes on the performance of plastic solar cells. Appl Phys Lett 80:1288

92. Brown TM, Millard IS, Lacey DJ, Burroughes JH, Friend RH, Cacialli F (2001) The influence of LiF thickness on the built-in potential of blue polymer light-emitting diodes with LiF/Al cathodes. Synth Met 124:15

93. van Gennip WJH, van Duren JKJ, Thüne PC, Janssen RAJ, Niemantsverdriet JW (2002) The interfaces of poly(p-phenylene vinylene) and fullerene derivatives with Al, LiF, and Al/LiF studied by secondary ion mass spectroscopy and X-ray photoelectron spectroscopy: formation of AI/F_3 disproved. J Chem Phys 117:5031

94. Deng XY, Tong SW, Hung LS, Mo YQ, Cao Y (2003) Role of ultrathin Alq3 and LiF layers in conjugated polymer light-emitting diodes. Appl Phys Lett 82:3104

95. Yokoyama T, Yoshimura D, Ito E, Ishii H, Ouchi Y, Seki K (2003) Energy level alignment at Alq3/LiF/Al interfaces studied by electron spectroscopies: island growth of LiF and size-dependence of the electronic structures. Jpn J Appl Phys 42:3666

96. Liu J, Duggal AR, Shiang JJ, Heller CM (2004) Efficient bottom cathodes for organic light-emitting devices. Appl Phys Lett 85:837

97. Jönsson SKM, Carlegrim E, Zhang F, Salaneck WR, Fahlman M (2005) Photoelectron spectroscopy of the contact between the cathode and the active layers in plastic solar cells: the role of LiF. Jpn J Appl Phys 44:3695

98. Gregg BA, Hanna MC (2003) Comparing organic to inorganic photovoltaic cells: theory, experiment, and simulation. J Appl Phys 93:3605

99. Kim H, Jin S-H, Suh H, Lee K (2004) Origin of the open circuit voltage in conjugated polymer–fullerene photovoltaic cells. In: Kafafi ZH, Lane PA (eds) Organic photovoltaics IV. SPIE Proc 5215:111

100. Gadisa A, Svensson M, Andersson MR, Inganäs O (2004) Correlation between oxidation potential and open-circuit voltage of composite solar cells based on blends of polythiophenes/fullerene derivative. Appl Phys Lett 84:1609

101. Hoppe H, Egbe DAM, Mühlbacher D, Sariciftci NS (2004) Photovoltaic action of conjugated polymer/fullerene bulk heterojunction solar cells using novel PPE-PPV copolymers. J Mater Chem 14:3461

102. Scharber MC, Mühlbacher D, Koppe M, Denk P, Waldauf C, Heeger AJ, Brabec CJ (2006) Design rules for donors in bulk-heterojunction solar cells—towards 10% energy-conversion efficiency. Adv Mater 18:789

103. Brabec CJ, Cravino A, Meissner D, Sariciftci NS, Fromherz T, Rispens MT, Sanchez L, Hummelen JC (2001) Origin of the open circuit voltage of plastic solar cells. Adv Funct Mater 11:374

104. Kooistra FB, Knol J, Kastenberg F, Popescu LM, Verhees WJH, Kroon JM, Hummelen JC (2007) Increasing the open circuit voltage of bulk-heterojunction solar cells by raising the LUMO level of the acceptor. Org Lett 9:551

105. Mihailetchi VD, Blom PWM, Hummelen JC, Rispens MT (2003) Cathode dependence of the open-circuit voltage of polymer: fullerene bulk heterojunction solar cells. J Appl Phys 94:6849

106. Heller CM, Campbell IH, Smith DL, Barashkov NN, Ferraris JP (1997) Chemical potential pinning due to equilibrium electron transfer at metal/C_{60}-doped polymer interfaces. J Appl Phys 81:3227

107. Hirose Y, Kahn A, Aristov V, Soukiassian P, Bulovic V, Forrest SR (1996) Chemistry and electronic properties of metal–organic semiconductor interfaces: Al, Ti, In, Sn, Ag, and Au on PTCDA. Phys Rev B 54:13748

108. Ishii H, Sugiyama K, Ito E, Seki K (1999) Energy level alignment and interfacial electronic structures at organic/metal and organic/organic interfaces. Adv Mater 11:605

109. Yan L, Gao Y (2002) Interfaces in organic semiconductor devices. Thin Solid Films 417:101

110. Koch N, Kahn A, Ghijsen J, Pireaux J-J, Schwartz J, Johnson RL, Elschner A (2003) Conjugated organic molecules on metal versus polymer electrodes: demonstration of a key energy level alignment mechanism. Appl Phys Lett 82:70

111. Cahen D, Kahn A (2003) Electron energetics at surfaces and interfaces: concepts and experiments. Adv Mater 15:271

112. Veenstra SC, Jonkman HT (2003) Energy-level alignment at metal–organic and organic–organic interfaces. J Polym Sci Polym Phys 41:2549

113. Veenstra SC, Heeres A, Hadziioannou G, Sawatzky GA, Jonkman HT (2002) On interface dipole layers between C_{60} and Ag or Au. Appl Phys A 75:661

114. Melzer C, Krasnikov VV, Hadziioannou G (2003) Organic donor/acceptor photovoltaics: the role of C_{60}/metal interfaces. Appl Phys Lett 82:3101

115. van Duren JKJ, Loos J, Morrissey F, Leewis CM, Kivits KPH, van IJzendoorn LJ, Rispens MT, Hummelen JC, Janssen RAJ (2002) In-situ compositional and structural analysis of plastic solar cells. Adv Funct Mater 12:665

116. Bulle-Lieuwma CWT, van Gennip WJH, van Duren JKJ, Jonkheijm P, Janssen RAJ, Niemantsverdriet JW (2003) Characterization of polymer solar cells by TOF-SIMS depth profiling. Appl Surf Sci 203–204:547

117. Gao J, Hide F, Wang H (1997) Efficient photodetectors and photovoltaic cells from composites of fullerenes and conjugated polymers: photoinduced electron transfer. Synth Met 84:979

118. Liu J, Shi Y, Yang Y (2001) Solvation-induced morphology effects on the performance of polymer-based photovoltaic devices. Adv Funct Mater 11:420

119. Scharber MC, Schulz NA, Sariciftci NS, Brabec CJ (2003) Optical- and photocurrent-detected magnetic resonance studies on conjugated polymer/fullerene composites. Phys Rev B 67:085202

120. van Duren JKJ, Yang X, Loos J, Bulle-Lieuwma CWT, Sieval AB, Hummelen JC, Janssen RAJ (2004) Relating the morphology of poly(p-phenylene vinylene)/methanofullerene blends to solar-cell performance. Adv Funct Mater 14:425

121. Riedel I, von Hauff E, Parisi J, Martin N, Giacalone F, Dyakonov V (2005) Dimethanofullerenes: new and efficient acceptors in bulk-heterojunction solar cells. Adv Funct Mater 15:1979

122. Katz EA, Faiman D, Tuladhar SM, Kroon JM, Wienk MM, Fromherz T, Padinger F, Brabec CJ, Sariciftci NS (2001) Temperature dependence for the photovoltaic device parameters of polymer–fullerene solar cells under operating conditions. J Appl Phys 90:5343

123. Dyakonov V (2002) The polymer–fullerene interpenetrating network: one route to a solar cell approach. Physica E 14:53

124. Riedel I, Parisi J, Dyakonov V, Lutsen L, Vanderzande D, Hummelen JC (2004) Effect of temperature and illumination on the electrical characteristics of polymer–fullerene bulk-heterojunction solar cells. Adv Funct Mater 14:38

125. Ramsdale CM, Barker JA, Arias AC, MacKenzie JD, Friend RH, Greenham NC (2002) The origin of the open circuit voltage in polyfluorene-based photovoltaic devices. J Appl Phys 92:4266

126. Peumans P, Yakimov A, Forrest SR (2003) Small molecular weight organic thin-film photodetectors and solar cells. J Appl Phys 93:3693

127. Schilinsky P, Waldauf C, Hauch J, Brabec CJ (2004) Simulation of light intensity dependent current characteristics of polymer solar cells. J Appl Phys 95:2816

128. Frohne H, Shaheen SE, Brabec CJ, Müller DC, Sariciftci NS, Meerholz K (2002) Influence of the anodic work function on the performance of organic solar cells. ChemPhysChem 9:795

129. Barker JA, Ramsdale CM, Greenham NC (2003) Modeling the current–voltage characteristics of bilayer polymer photovoltaic devices. Phys Rev B 67:075205

130. Hummelen JC, Knight BW, LePeq F, Wudl F, Yao J, Wilkins CL (1995) Preparation and characterization of fulleroid and methanofullerene derivatives. J Org Chem 60:532

131. Yang CY, Heeger AJ (1996) Morphology of composites of semiconducting polymers mixed with C_{60}. Synth Met 83:85

132. Drees M, Premaratne K, Graupner W, Heflin JR, Davis RM, Marciu D, Miller M (2002) Creation of a gradient polymer–fullerene interface in photovoltaic devices by thermally controlled interdiffusion. Appl Phys Lett 81:1

133. Drees M, Davis RM, Heflin JR (2005) Improved morphology of polymer–fullerene photovoltaic devices with thermally induced concentration gradients. J Appl Phys 97:036103

134. Martens T, D'Haen J, Munters T, Goris L, Beelen Z, Manca J, D'Olieslaeger M, Vanderzande D, Schepper LD, Andriessen R (2002) The influence of the microstructure upon the photovoltaic performance of MDMOPPV:PCBM bulk hetero-junction organic solar cells. Presented at the Materials Research Society Spring Meeting, San Francisco. MRS Proc 725:P7111

135. Martens T, D'Haen J, Munters T, Beelen Z, Goris L, Manca J, D'Olieslaeger M, Vanderzande D, Schepper LD, Andriessen R (2003) Disclosure of the nanostructure of MDMO-PPV:PCBM bulk heterojunction organic solar cells by a combination of SPM and TEM. Synth Met 138:243

136. Martens T, Beelen Z, D'Haen J, Munters T, Goris L, Manca J, D'Olieslaeger M, Vanderzande D, Schepper LD, Andriessen R (2003) Morphology of MDMO-PPV:PCBM bulk hetero-junction organic solar cells studied by AFM, KFM and TEM. In: Kafafi Z H, Fichou D (eds) Organic photovoltaics III. SPIE Proc 4801:40

137. Hoppe H, Drees M, Schwinger W, Schäffler F, Sariciftci NS (2005) Nano-crystalline fullerene phases in polymer/fullerene bulk-heterojunction solar cells: a transmission electron microscopy study. Synth Met 152:117

138. McNeill CR, Frohne H, Holdsworth JL, Dastoor PC (2004) Direct influence of morphology on current generation in conjugated polymer:methanofullerene solar cells measured by near-field scanning photocurrent microscopy. Synth Met 147:101

139. Geens W, Shaheen SE, Brabec CJ, Poortmans J, Sariciftci NS (2000) Field-effect mobility measurements of conjugated polymer/fullerene photovoltaic blends. Presented at the electronic properties of novel materials—molecular nanostructures, 14th international winter school/Euroconference (American Institute of Physics), Kirchberg

140. Geens W, Shaheen SE, Wessling B, Brabec CJ, Poortmans J, Sariciftci NS (2002) Dependence of field-effect hole mobility of PPV-based polymer films on the spin-casting solvent. Org Electron 3:105

141. Aernouts T, Vanlaeke P, Geens W, Poortmans J, Heremans P, Borghs S, Mertens R (2003) The influence of the donor/acceptor ratio on the performance of organic bulk heterojunction solar cells. Presented at the E-MRS spring meeting, Strasbourg

142. Pacios R, Nelson J, Bradley DDC, Brabec CJ (2003) Composition dependence of electron and hole transport in polyfluorene:[6,6]-phenyl C61-butyric acid methyl ester blend films. Appl Phys Lett 83:4764

143. Melzer C, Koop EJ, Mihailetchi VD, Blom PWM (2004) Hole transport in poly(phenylene vinylene)/methanofullerene bulk-heterojunction solar cells. Adv Funct Mater 14:865

144. Mihailetchi VD, Koster LJA, Blom PWM, Melzer C, de Boer B, van Duren JKJ, Janssen RAJ (2005) Compositional dependence of the performance of poly(p-phenylene vinylene):methanofullerene bulk-heterojunction solar cells. Adv Funct Mater 15:795

145. Dennler G, Mozer AJ, Juska G, Pivrikas A, Österbacka R, Fuchsbauer A, Sariciftci NS (2006) Charge carrier mobility and lifetime versus composition of conjugated polymer/fullerene bulk-heterojunction solar cells. Org Electron 7:229

146. Gadisa A, Wang X, Admassie S, Perzon E, Oswald F, Langa F, Andersson MR, Inganäs O (2006) Stoichiometry dependence of charge transport in polymer/methanofullerene and polymer/C_{70} derivative based solar cells. Org Electron 7:195

147. Andersson LM, Inganäs O (2006) Acceptor influence on hole mobility in fullerene blends with alternating copolymers of fluorene. Appl Phys Lett 88:082103

148. Mihailetchi VD, Wildeman J, Blom PWM (2005) Space-charge limited photocurrent. Phys Rev Lett 94:126602

149. Bässler H (1993) Charge transport in disordered organic photoconductors. Phys Status Solidi B 175:15

150. Dyakonov V (2004) Mechanisms controlling the efficiency of polymer solar cells. Appl Phys A 79:21

151. Riedel I, Dyakonov V (2004) Influence of electronic transport properties of polymer–fullerene blends on the performance of bulk heterojunction photovoltaic devices. Phys Status Solidi A 201:1332

152. Onsager L (1934) Deviations from Ohm's law in weak electrolytes. J Chem Phys 2:599

153. Braun CL (1984) Electric field assisted dissociation of charge transfer states as a mechanism of photocarrier production. J Chem Phys 80:4157

154. Gommans HHP, Kemerink M, Kramer JM, Janssen RAJ (2005) Field and temperature dependence of the photocurrent in polymer/fullerene bulk heterojunction solar cells. Appl Phys Lett 87:122104

155. Koster LJA, Mihailetchi VD, Ramaker R, Blom PWM (2005) Light intensity dependence of open-circuit voltage of polymer:fullerene solar cells. Appl Phys Lett 86:123509

156. Koster LJA, Mihailetchi VD, Xie H, Blom PWM (2005) Origin of the light intensity dependence of the short-circuit current of polymer/fullerene solar cells. Appl Phys Lett 87:203502

157. Waldauf C, Schilinsky P, Hauch J, Brabec CJ (2004) Material and device concepts for organic photovoltaics: towards competitive efficiencies. Thin Solid Films 451–452:503

158. Waldauf C, Scharber MC, Schilinsky P, Hauch JA, Brabec CJ (2006) Physics of organic bulk heterojunction devices for photovoltaic applications. J Appl Phys 99:104503

159. Harbecke B (1986) Coherent and incoherent reflection and transmission of multilayer systems. Appl Phys B 39:165

160. Rostalski J, Meissner D (2000) Photocurrent spectroscopy for the investigation of charge carrier generation and transport mechanisms in organic p/n-junction solar cells. Sol Energy Mater Sol Cells 63:37
161. Gruber DP, Meinhardt G, Papousek W (2005) Modelling the light absorption in organic photovoltaic devices. Sol Energy Mater Sol Cells 87:215
162. Niggemann M, Bläsi B, Gombert A, Hinsch A, Hoppe H, Lalanne P, Meissner D, Wittwer V (2002) Trapping light in organic plastic solar cells with integrated diffraction gratings. Presented at the 17th European photovoltaic solar energy conference, Munich, 22–26 October 2001
163. Hoppe H, Arnold N, Meissner D, Sariciftci NS (2004) Modeling of optical absorption in conjugated polymer/fullerene bulk-heterojunction plastic solar cells. Thin Solid Films 451–452:589
164. Persson N-K, Schubert M, Inganäs O (2004) Optical modelling of a layered photovoltaic device with a polyfluorene derivative/fullerene as the active layer. Sol Energy Mater Sol Cells 83:169
165. Hoppe H, Shokhovets S, Gobsch G (2007) Inverse relation between photocurrent and absorption layer thickness in polymer solar cells. Phys Status Solidi RRL 1:R40
166. Slooff LH, Veenstra SC, Kroon JM, Moet DJD, Sweelssen J, Koetse MM (2007) Determining the internal quantum efficiency of highly efficient polymer solar cells through optical modeling. Appl Phys Lett 90:143506
167. Lutsen L, Adriaensens P, Becker H, van Breemen AJ, Vanderzande D, Gelan J (1999) New synthesis of a soluble high molecular weight poly(arylene vinylene): poly[2-methoxy-5-(3,7-dimethyloctyloxy)-p-phenylene vinylene] polymerization and device properties. Macromolecules 32:6517
168. Munters T, Martens T, Goris L, Vrindts V, Manca J, Lutsen L, Ceunick WD, Vanderzande D, Schepper LD, Gelan J, Sariciftci NS, Brabec CJ (2002) A comparison between state-of-the-art gilch and sulphinyl synthesised MDMO-PPV/PCBM bulk hetero-junction solar cells. Thin Solid Films 403–404:247
169. Mozer A, Denk P, Scharber M, Neugebauer H, Sariciftci NS, Wagner P, Lutsen L, Vanderzande D (2004) Novel regiospecific MDMO-PPV copolymer with improved charge transport for bulk heterojunction solar cells. J Phys Chem B 108:5235
170. Wienk MM, Kroon JM, Verhees WJH, Knol J, Hummelen JC, van Hall PA, Janssen RAJ (2003) Efficient methano[70]fullerene/MDMO-PPV bulk heterojunction photovoltaic cells. Angew Chem Int Ed 42:3371
171. Camaioni N, Ridolfi G, Casalbore-Miceli G, Possamai G, Maggini M (2002) The effect of a mild thermal treatment on the performance of poly(3-alkylthiophene)/fullerene solar cells. Adv Mater 14:1735
172. Padinger F, Rittberger RS, Sariciftci NS (2003) Effects of postproduction treatment on plastic solar cells. Adv Funct Mater 13:1
173. Zhao Y, Yuan GX, Roche P, Leclerc M (1995) A calorimetric study of the phase transitions in poly(3-hexylthiophene). Polymer 36:2211
174. Berggren M, Gustafsson G, Inganäs O, Andersson MR, Wennerström O, Hjertberg T (1994) Thermal control of near-infrared and visible electroluminescence in alkyl-phenyl substituted polythiophenes. Appl Phys Lett 65:1489
175. Chirvase D, Parisi J, Hummelen JC, Dyakonov V (2004) Influence of nanomorphology on the photovoltaic action of polymer–fullerene composites. Nanotechnology 15:1317
176. Brown PJ, Thomas DS, Köhler A, Wilson J, Kim JS, Ramsdale C, Sirringhaus H, Friend RH (2003) Effect of interchain interactions on the absorption and emission of poly(3-hexylthiophene). Phys Rev B 67:064203

177. Ahn T, Lee H, Han S-H (2002) Effect of annealing of polythiophene derivative for polymer light-emitting diodes. Appl Phys Lett 80:392
178. Kim Y, Choulis SA, Nelson J, Bradley DDC, Cook S, Durrant JR (2005) Device annealing effect in organic solar cells with blends of regioregular poly(3-hexylthiophene) and soluble fullerene. Appl Phys Lett 86:063502
179. Yang X, Loos J, Veenstra SC, Verhees WJH, Wienk MM, Kroon JM, Michels MAJ, Janssen RAJ (2005) Nanoscale morphology of high-performance polymer solar cells. Nano Lett 5:579
180. Ihn KJ, Moulton J, Smith P (1993) Whiskers of poly(3-alkylthiophene)s. J Polym Sci B Polym Phys 31:735
181. Erb T, Zhokhavets U, Gobsch G, Raleva S, Stühn B, Schilinsky P, Waldauf C, Brabec CJ (2005) Correlation between structural and optical properties of composite polymer films for organic solar cells. Adv Funct Mater 15:1193
182. Erb T, Zhokhavets U, Hoppe H, Gobsch G, Al-Ibrahim M, Ambacher O (2006) Absorption and crystallinity of poly(3-hexylthiophene)/fullerene blends in dependence on annealing temperature. Thin Solid Films 511–512:483
183. Zhokhavets U, Erb T, Gobsch G, Al-Ibrahim M, Ambacher O (2006) Relation between absorption and crystallinity of poly(3-hexylthiophene)/fullerene films for plastic solar cells. Chem Phys Lett 418:343
184. Yang X, van Duren JKJ, Rispens MT, Hummelen JC, Janssen RAJ, Michels MAJ, Loos J (2004) Crystalline organization of a methanofullerene as used for plastic solar-cell applications. Adv Mater 16:802
185. Schuller S, Schilinsky P, Hauch J, Brabec CJ (2004) Determination of the degradation constant of bulk heterojunction solar cells by accelerated lifetime measurements. Appl Phys A Mater Sci Proc 79:37
186. Drees M, Hoppe H, Winder C, Neugebauer H, Sariciftci NS, Schwinger W, Schäffler F, Topf C, Scharber MC, Zhu Z, Gaudiana R (2005) Stabilization of the nanomorphology of polymer/fullerene bulk heterojunction blends using a novel polymerizable fullerene derivative. J Mater Chem 15:5158
187. Sivula K, Ball ZT, Watanabe N, Fréchet JMJ (2006) Amphiphilic diblock copolymer compatibilizers and their effect on the morphology and performance of polythiophene:fullerene solar cells. Adv Mater 18:206
188. Brabec CJ, Hauch JA, Schilinsky P, Waldauf C (2005) Production aspects of organic photovoltaics and their impact on the commercialization of devices. MRS Bull 30:50
189. Huang J, Li G, Yang Y (2005) Influence of composition and heat-treatment on the charge transport properties of poly(3-hexylthiophene) and [6,6]-phenyl C_{61}-butyric acid methyl ester blends. Appl Phys Lett 87:112105
190. Kim Y, Cook S, Tuladhar SM, Choulis SA, Nelson J, Durrant JR, Bradley DDC, Giles M, McCulloch I, Ha C-S, Ree M (2006) A strong regioregularity effect in self-organizing conjugated polymer films and high-efficiency polythiophene:fullerene solar cells. Nat Mater 5:197
191. Li G, Shrotriya V, Yao Y, Yang Y (2005) Investigation of annealing effects and film thickness dependence of polymer solar cells based on poly(3-hexylthiophene). J Appl Phys 98:043704
192. Moulé AJ, Bonekamp JB, Meerholz K (2006) The effect of active layer thickness and composition on the performance of bulk-heterojunction solar cells. J Appl Phys 100:094503
193. Schilinsky P, Asawapirom U, Scherf U, Biele M, Brabec CJ (2005) Influence of the molecular weight of poly(3-hexylthiophene) on the performance of bulk heterojunction solar cells. Chem Mater 17:2175

194. Schilinsky P, Waldauf C, Brabec CJ (2002) Recombination and loss analysis in poly-thiophene based bulk heterojunction photodetectors. Appl Phys Lett 81:3885

195. Svensson M, Zhang F, Veenstra SC, Verhees WJH, Hummelen JC, Kroon JM, Inganäs O, Andersson MR (2003) High-performance polymer solar cells of an alternating polyfluorene copolymer and a fullerene derivative. Adv Mater 15:988

196. Yohannes T, Zhang F, Svensson M, Hummelen JC, Andersson MR, Inganäs O (2004) Polyfluorene copolymer based bulk heterojunction solar cells. Thin Solid Films 449:152

197. Zhang F, Jespersen KG, Björström C, Svensson M, Andersson MR, Sundström V, Magnusson K, Moons E, Yartsev A, Inganäs O (2006) Influence of solvent mixing on the morphology and performance of solar cells based on polyfluorene copoly-mer/fullerene blends. Adv Funct Mater 16:667

198. Roncali J (1997) Synthetic principles for band gap control in linear π-conjugated systems. Chem Rev 97:173

199. Winder C, Sariciftci NS (2004) Low bandgap polymers for photon harvesting in bulk heterojunction solar cells. J Mater Chem 14:1077

200. Thompson BC, Kim Y-G, Reynolds JR (2005) Spectral broadening in MEH-PPV:PCBM-based photovoltaic devices via blending with narrow band gap cyanovinylene-dioxythiophene polymer. Macromolecules 38:5359

201. Dhanabalan A, Knol J, Hummelen JC, Janssen RAJ (2001) Design and synthesis of new processible donor–acceptor dyad and triads. Synth Met 119:519

202. van Duren JKJ, Dhanabalan A, van Hal PA, Janssen RAJ (2001) Low-bandgap poly-mer photovoltaic cells. Synth Met 121:1587

203. Brabec CJ, Winder C, Sariciftci NS, Hummelen JC, Dhanabalan A, van Hal PA, Janssen RAJ (2002) A low-bandgap semiconducting polymer for photovoltaic de-vices and infrared emitting diodes. Adv Funct Mater 12:709

204. Colladet K, Nicolas M, Goris L, Lutsen L, Vanderzande D (2004) Low-band gap polymers for photovoltaic applications. Thin Solid Films 451–452:7

205. Zhou Q, Hou Q, Zheng L, Deng X, Yu G, Cao Y (2004) Fluorene-based low band-gap copolymers for high performance. Appl Phys Lett 84:1653

206. Campos LM, Tontcheva A, Günes S, Sonmez G, Neugebauer H, Sariciftci NS, Wudl F (2005) Extended photocurrent spectrum of a low band gap polymer in a bulk het-erojunction solar cell. Chem Mater 17:4031

207. Wang X, Perzon E, Delgado JL, de la Cruz P, Zhang F, Langa F, Andersson M, In-ganäs O (2004) Infrared photocurrent spectral response from plastic solar cell with low-bandgap polyfluorene and fullerene derivative. Appl Phys Lett 85:5081

208. Wang X, Perzon E, Oswald F, Langa F, Admassie S, Andersson MR, Inganäs O (2005) Enhanced photocurrent spectral response in low-bandgap polyfluorene and C_{70}-derivative-based solar cells. Adv Funct Mater 15:1665

209. Perzon E, Wang X, Zhang F, Mammo W, Delgado JL, de la Cruz P, Inganäs O, Langa F, Andersson MR (2005) Design, synthesis and properties of low-bandgap polyfluorenes for photovoltaic devices. Synth Met 154:53

210. Wang X, Perzon E, Mammo W, Oswald F, Admassie S, Persson N-K, Langa F, Ander-sson MR, Inganäs O (2006) Polymer solar cells with low-bandgap polymers blended with C_{70}-derivative give photocurrent at 1 μm. Thin Solid Films 511–512:576

211. Perzon E, Wang X, Admassie S, Inganäs O, Andersson MR (2006) An alternating low band-gap polyfluorene for optoelectronic devices. Polymer 47:4261

212. Zhang F, Perzon E, Wang X, Mammo W, Andersson MR, Inganäs O (2005) Polymer solar cells based on a low-bandgap fluorene copolymer and a fullerene derivative with photocurrent extended to 850 nm. Adv Funct Mater 15:745

213. Zhang F, Mammo W, Andersson LM, Admassie S, Andersson MR, Inganäs O (2006) Low-bandgap alternating fluorene copolymer/methanofullerene heterojunctions in efficient near-infrared polymer solar cells. Adv Mater 18:2169

214. Lee SK, Cho NS, Kwak JH, Lim KS, Shim H-K, Hwang D-H, Brabec CJ (2006) New low band-gap alternating polyfluorene derivatives for photovoltaic cells. Thin Solid Films 511–512:157

215. Wienk M, Struijk MP, Janssen RAJ (2006) Low bandgap polymer bulk heterojunction solar cells. Chem Phys Lett 422:488

216. Iou J, Hou Q, Chen J, Cao Y (2006) Luminescence and photovoltaic cells of benzo-selenadiazole-containing polyfluorenes. Synth Met 156:470

217. Wienk MM, Turbiez MGR, Struijk MP, Fonrodona M, Janssen RAJ (2006) Low-band gap poly(di-2-thienylthienopyrazine):fullerene solar cells. Appl Phys Lett 88:153511

218. Halls JJM, Friend RH (1997) The photovoltaic effect in a poly(p-phenylenevinylene)/perylene heterojunction. Synth Met 85:1307

219. Dittmer JJ, Lazzaroni R, Leclere P, Moretti P, Granström M, Petritsch K, Marseglia EA, Friend RH, Bredas JL, Rost H, Holmes AB (2000) Crystal network formation in organic solar cells. Sol Energy Mater Sol Cells 61:53

220. Deng X, Zheng L, Yang C, Li Y, Yu G, Cao Y (2004) Polymer photovoltaic devices fabricated with blend MEHPPV and organic small molecules. J Phys Chem B 108:3451

221. Halls JJM, Cornil J, de Santos DA, Silbey R, Hwang D-H, Holmes AB, Brédas JL, Friend RH (1999) Charge- and energy-transfer processes at polymer/polymer interfaces: a joint experimental and theoretical study. Phys Rev B 60:5721

222. Alam MM, Jenekhe SA (2004) Efficient solar cells from layered nanostructures of donor and acceptor conjugated polymers. Chem Mater 16:4647

223. Breeze AJ, Schlesinger Z, Carter SA, Tillmann H, Hörhold H-H (2004) Improving power efficiencies in polymer–polymer blend photovoltaics. Sol Energy Mater Sol Cells 83:263

224. Chasteen SV, Härter JO, Rumbles G, Scott JC, Nakazawa Y, Jones M, Hörhold H-H, Tillman H, Carter SA (2006) Comparison of blended versus layered structures for poly(p-phenylene vinylene)-based polymer photovoltaics. J Appl Phys 99:033709

225. Kietzke T, Egbe DAM, Hörhold H-H, Neher D (2006) Comparative study of M3EH-PPV-based bilayer photovoltaic devices. Macromolecules 39:4018

226. Kietzke T, Hörhold H-H, Neher D (2005) Efficient polymer solar cells based on M3EH-PPV. Chem Mater 17:6532

227. Veenstra SC, Verhees WJH, Kroon JM, Koetse MM, Sweelssen J, Bastiaansen JJAM, Schoo HFM, Yang X, Alexeev A, Loos J, Schubert US, Wienk MM (2004) Photovoltaic properties of a conjugated polymer blend of MDMO-PPV and PCNEPV. Chem Mater 16:2503

228. Quist PAC, Savenije TJ, Koetse MM, Veenstra SC, Kroon JM, Siebbeles LDA (2005) The effect of annealing on the charge-carrier dynamics in a polymer/polymer bulk heterojunction for photovoltaic applications. Adv Funct Mater 15:469

229. Offermans T, van Hal PA, Meskers SCJ, Koetse MM, Janssen RAJ (2005) Exciplex dynamics in a blend of p-conjugated polymers with electron donating and accepting properties: MDMO-PPV and PCNEPV. Phys Rev B 72:045213

230. Veldman D, Offermans T, Sweelssen J, Koetse MM, Meskers SCJ, Janssen RAJ (2006) Triplet formation from the charge-separated state in blends of MDMO-PPV with cyano-containing acceptor polymers. Thin Solid Films 511–512:333

231. Russell DM, Arias AC, Friend RH, Silva C, Ego C, Grimsdale AC, Müllen K (2002) Efficient light harvesting in a photovoltaic diode composed of a semiconductor conjugated copolymer blend. Appl Phys Lett 80:2204

232. Kim Y, Cook S, Choulis SA, Nelson J, Durrant JR, Bradley DDC (2004) Organic photovoltaic devices based on blends of regioregular poly(3-hexylthiophene) and poly(9,9-dioctylfluorene-co-benzothiadiazole). Chem Mater 16:4812

233. Ridolfi G, Camaioni N, Samori P, Gazzano M, Accorsi G, Armaroli N, Favaretto L, Barbarella G (2005) All-thiophene donor–acceptor blends: photophysics, morphology and photoresponse. J Mater Chem 15:895

234. Koetse MM, Sweelssen J, Hoekerd KT, Schoo HFM, Veenstra SC, Kroon JM, Yang X, Loos J (2006) Efficient polymer:polymer bulk heterojunction solar cells. Appl Phys Lett 88:083504

235. Halls JJM, Arias AC, MacKenzie JD, Wu W, Inbasekaran M, Woo EP, Friend RH (2000) Photodiodes based on polyfluorene composites: influence of morphology. Adv Mater 12:498

236. Arias AC, MacKenzie JD, Stevenson R, Halls JJM, Inbasekaran M, Woo EP, Richards D, Friend RH (2001) Photovoltaic performance and morphology of polyfluorene blends: a combined microscopic and photovoltaic investigation. Macromolecules 34:6005

237. Snaith HJ, Arias AC, Morteani AC, Silva C, Friend RH (2002) Charge generation kinetics and transport mechanisms in blended polyfluorene photovoltaic devices. Nano Lett 2:1353

238. Xia Y, Friend RH (2005) Controlled phase separation of polyfluorene blends via inkjet printing. Macromolecules 38:6466

239. Kietzke T, Neher D, Landfester K, Montenegro R, Güntner R, Scherf U (2003) Novel approaches to polymer blends based on polymer nanoparticles. Nat Mater 2:408

240. Kietzke T, Neher D, Kumke M, Montenegro R, Landfester K, Scherf U (2004) A nanoparticle approach to control the phase separation in polyfluorene photovoltaic devices. Macromolecules 37:4882

241. Arias AC, Corcoran N, Banach M, Friend RH, MacKenzie JD, Huck WTS (2002) Vertically segregated polymer-blend photovoltaic thin-film structures through surface-mediated solution processing. Appl Phys Lett 80:1695

242. Pacios R, Bradley DDC (2002) Charge separation in polyfluorene composites with internal donor/acceptor heterojunctions. Synth Met 127:261

243. Chiesa M, Bürgi L, Kim J-S, Shikler R, Friend RH, Sirringhaus H (2005) Correlation between surface photovoltage and blend morphology in polyfluorene-based photodiodes. Nano Lett 5:559

244. Glatzel T, Hoppe H, Sariciftci NS, Lux-Steiner MC, Komiyama M (2005) Kelvin probe force microscopy study on conjugated polymer/fullerene organic solar cells. Jpn J Appl Phys 44:5370

245. Park Y, Choong V, Gao Y, Hsieh BR, Tang CW (1996) Work funtion of indium tin oxide transparent conductor measured by photoelectron spectroscopy. Appl Phys Lett 68:2699

246. Snaith HJ, Greenham NC, Friend RH (2004) The origin of collected charge and open-circuit voltage in blended polyfluorene photovoltaic devices. Adv Mater 16:1640

247. Onsager L (1938) Initial recombination of ions. Phys Rev 54:554

248. Dhoot AS, Hogan JA, Morteani AC, Greenham NC (2004) Electromodulation of photoinduced charge transfer in polyfluorene bilayer devices. Appl Phys Lett 85:2256

249. Greenham NC, Peng X, Alivisatos AP (1996) Charge separation and transport in conjugated-polymer/semiconductor-nanocrystal composites studied by photoluminescence quenching and photoconductivity. Phys Rev B 54:17628

250. Huynh WU, Dittmer JJ, Alivisato AP (2002) Hybrid nanorod–polymer solar cells. Science 295:2425

251. Huynh WU, Dittmer JJ, Libby WC, Whiting GL, Alivisato AP (2003) Controlling the morphology of nanocrystal–polymer composites for solar cells. Adv Funct Mater 13:73

252. Pientka M, Dyakonov V, Meissner D, Rogach A, Talapin D, Weller H, Lutsen L, Vanderzande D (2004) Photoinduced charge transfer in composites of conjugated polymers and semiconductor nanocrystals. Nanotechnology 15:163

253. Pientka M, Wisch J, Böger S, Parisi J, Dyakonov V, Rogach A, Talapin D, Weller H (2004) Photogeneration of charge carriers in blends of conjugated polymers and semiconducting nanoparticles. Thin Solid Films 451–452:48

254. Sun B, Marx E, Greenham NC (2003) Photovoltaic devices using blends of branched CdSe nanoparticles and conjugated polymers. Nano Lett 3:961

255. Sun B, Snaith HJ, Dhoot AS, Westenhoff S, Greenham NC (2005) Vertically segregated hybrid blends for photovoltaic devices with improved efficiency. J Appl Phys 97:014914

256. Snaith HJ, Whiting GL, Sun B, Greenham NC, Huck WTS, Friend RH (2005) Self-organization of nanocrystals in polymer brushes: application in heterojunction photovoltaic diodes. Nano Lett 5:1653

257. Lin Y, Böker A, He J, Sill K, Xiang H, Abetz C, Li X, Wang J, Emrick T, Long S, Wang Q, Balazs A, Russell TP (2005) Self-directed self-assembly of nanoparticle/copolymer mixtures. Nature 434:55

258. Firth AV, Tao Y, Wang D, Ding J, Bensebaa F (2005) Microwave assisted synthesis of CdSe nanocrystals for straightforward integration into composite photovoltaic devices. J Mater Chem 15:4367

259. Landi BJ, Castro SL, Ruf HJ, Evans CM, Bailey SG, Raffaelle RP (2005) CdSe quantum dot–single wall carbon nanotube complexes for polymeric solar cells. Sol Energy Mater Sol Cells 87:733

260. Liang Z, Dzienis KL, Xu J, Wang Q (2006) Covalent layer-by-layer assembly of conjugated polymers and CdSe nanoparticles: multilayer structure and photovoltaic properties. Adv Funct Mater 16:542

261. Kang Y, Kim D (2006) Well-aligned CdS nanorod/conjugated polymer solar cells. Sol Energy Mater Sol Cells 90:166

262. Kannan B, Castelino K, Majumdar A (2003) Design of nanostructured heterojunction polymer photovoltaic devices. Nano Lett 3:1729

263. Arici E, Sariciftci NS, Meissner D (2002) Photovoltaic properties of nanocrystalline CuInS$_2$/methanofullerene solar cells. Mol Cryst Liq Cryst 385:129

264. Arici E, Hoppe H, Reuning A, Sariciftci NS, Meissner D (2002) CIS plastic solar cells. Presented at the 17th European photovoltaic solar energy conference, Munich, 22–26 October 2001, p 61

265. Arici E, Sariciftci NS, Meissner D (2003) Hybrid solar cells based on nanoparticles of CuInS$_2$ in organic matrices. Adv Funct Mater 13:165

266. Arici E, Hoppe H, Schäffler F, Meissner D, Malik MA, Sariciftci NS (2004) Hybrid solar cells based on inorganic nanoclusters and semiconductive polymers. Thin Solid Films 451–452:612

267. Arici E, Hoppe H, Schäffler F, Meissner D, Malik MA, Sariciftci NS (2004) Morphology effects in nanocrystalline CuInSe$_2$-conjugated polymer hybrid systems. Appl Phys A 79:59

268. Kang Y, Park N-G, Kim D (2005) Hybrid solar cells with vertically aligned CdTe nanorods and a conjugated polymer. Appl Phys Lett 86:113101

269. Watt AAR, Blake D, Warner JH, Thomson EA, Tavenner EL, Rubinsztein-Dunlop H, Meredith P (2005) Lead sulfide nanocrystal:conducting polymer solar cells. J Phys D Appl Phys 38:2006

270. Zhang S, Cyr PW, McDonald SA, Konstantatos G, Sargent EH (2005) Enhanced infrared photovoltaic efficiency in PbS nanocrystal/semiconducting polymer composites: 600-fold increase in maximum power output via control of the ligand barrier. Appl Phys Lett 87:233101

271. Cui D, Xu J, Zhu T, Paradee G, Ashok S, Gerhold M (2006) Harvest of near infrared light in PbSe nanocrystal–polymer hybrid photovoltaic cells. Appl Phys Lett 88:183111

272. Günes S, Neugebauer H, Sariciftci NS, Roither J, Kovalenko M, Pillwein G, Heiss W (2006) Hybrid solar cells using HgTe nanocrystals and nanoporous TiO$_2$ electrodes. Adv Funct Mater 16:1095

273. Beek WJE, Wienk MM, Janssen RAJ (2004) Efficient hybrid solar cells from zinc oxide nanoparticles and a conjugated polymer. Adv Mater 16:1009

274. Beek WJE, Slooff LH, Wienk MM, Kroon JM, Janssen RAJ (2005) Hybrid solar cells using a zinc oxide precursor and a conjugated polymer. Adv Funct Mater 15:1703

275. Beek WJE, Wienk MM, Kemerink M, Yang X, Janssen RAJ (2005) Hybrid zinc oxide conjugated polymer bulk heterojunction solar cells. J Phys Chem B 109:9505

276. Beek WJE, Wienk MM, Janssen RAJ (2005) Hybrid polymer solar cells based on zinc oxide. J Mater Chem 15:2985

277. Beek WJE, Wienk MM, Janssen RAJ (2006) Hybrid solar cells from regioregular polythiophene and ZnO nanoparticles. Adv Funct Mater 16:1112

278. Greene LE, Law M, Tan DH, Montano M, Goldberger J, Somorjai G, Yang P (2005) General route to vertical ZnO nanowire arrays using textured ZnO seeds. Nano Lett 5:1231

279. Quist PAC, Slooff LH, Donker H, Kroon JM, Savanije TJ, Siebbeles LDA (2005) Formation and decay of charge carriers in hybrid MDMO-PPV:ZnO bulk heterojunctions produced from a ZnO precursor. Superlattices Microstruct 38:308

280. Olson DC, Piris J, Collins RT, Shaheen SE, Ginley DS (2006) Hybrid photovoltaic devices of polymer and ZnO nanofiber composites. Thin Solid Films 496:26

281. Peiro AM, Ravirajan P, Govender K, Boyle DS, O'Brien P, Bradley DDC, Nelson J, Durrant JR (2006) Hybrid polymer/metal oxide solar cells based on ZnO columnar structures. J Mater Chem 16:2088

282. Ravirajan P, Peiro AM, Nazeeruddin MK, Graetzel M, Bradley DDC, Durrant JR, Nelson J (2006) Hybrid polymer/zinc oxide photovoltaic devices with vertically oriented ZnO nanorods and an amphiphilic molecular interface layer. J Phys Chem B 110:7635

283. O'Regan B, Grätzel M (1991) A low cost, high-efficiency solar cell based on dye-sensitized colloidal TiO$_2$ films. Nature 353:737

284. Grätzel M (2001) Photoelectrochemical cells. Nature 414:338

285. Bach U, Lupo D, Comte P, Moser JE, Weissörtel F, Salbeck J, Spreitzer H, Grätzel M (1998) Solid-state dye-sensitized mesoporous TiO$_2$ solar cells with high photon-to-electron conversion efficiencies. Nature 395:583

286. Krüger J, Bach U, Grätzel M (2000) Modification of TiO$_2$ heterojunctions with benzoic acid derivatives in hybrid molecular solid-state devices. Adv Mater 12:447

287. Krüger J, Plass R, Cevey L, Piccirelli M, Grätzel M, Bach U (2001) High efficiency solid-state photovoltaic device due to inhibition of interface charge recombination. Appl Phys Lett 79:2085

288. Krüger J, Plass R, Grätzel M, Matthieu H-J (2002) Improvement of the photovoltaic performance of solid-state dye-sensitized device by silver complexation of the sensitizer *cis*-bis(4,4-dicarboxy-2,2-bipyridine)-bis(isothiocyanato)ruthenium(II). Appl Phys Lett 81:367

289. Kim Y-G, Walker J, Samuelson LA, Kumar J (2003) Efficient light harvesting polymers for nanocrystalline TiO_2 photovoltaic cells. Nano Lett 3:523

290. Senadeera GKR, Nakamura K, Kitamura T, Wada Y, Yanagida S (2003) Fabrication of highly efficient polythiophene-sensitized metal oxide photovoltaic cells. Appl Phys Lett 83:5470

291. Nogueira AF, Longo C, De Paoli M-A (2004) Polymers in dye sensitized solar cells: overview and perspectives. Coord Chem Rev 248:1455

292. Arango AC, Carter SA, Brock PJ (1999) Charge transfer in photovoltaics consisting of interpenetrating networks of conjugated polymer and TiO_2 nanoparticles. Appl Phys Lett 74:1698

293. Arango AC, Johnson LR, Bliznyuk VN, Schlesinger Z, Carter SA, Hörhold H-H (2000) Efficient titanium oxide/conjugated polymer photovoltaics for solar energy conversion. Adv Mater 12:1689

294. Salafsky JS (1999) Exciton dissociation, charge transport, and recombination in ultrathin, conjugated polymer–TiO_2 nanocrystal intermixed composites. Phys Rev B 59:10885

295. Breeze AJ, Schlesinger Z, Carter SA (2001) Charge transport in TiO_2/MEH-PPV polymer photovoltaics. Phys Rev B 64:125205

296. Gebeyehu D, Brabec CJ, Padinger F, Fromherz T, Spiekermann S, Vlachopoulos N, Kienberger F, Schindler H, Sariciftci NS (2001) Solid state dye-sensitized TiO_2 solar cells with poly(3-octylthiophene) as hole transport layer. Synth Met 121:1549

297. Kaneko M, Takayama K, Pandey SS, Takashima W, Endo T, Rikukawa M, Kaneto K (2001) Photovoltaic cell using high mobility poly(alkylthiophene)s and TiO_2. Synth Met 121:1537

298. Fan Q, McQuillin B, Bradley DDC, Whitelegg S, Seddon AB (2001) A solid-state solar cell using sol–gel processed material and a polymer. Chem Phys Lett 347:325

299. Qiao Q, McLeskey JJT (2005) Water-soluble polythiophene/nanocrystalline TiO_2 solar cells. Appl Phys Lett 86:153501

300. Coakley KM, Srinivasan BS, Ziebarth JM, Goh C, Liu Y, McGehee MD (2005) Enhanced hole mobility in regioregular polythiophene infiltrated in straight nanopores. Adv Funct Mater 15:1927

301. Bartholomew GP, Heeger AJ (2005) Infiltration of regioregular poly[2,2'-(3-hexylthiophene)] into random nanocrystalline TiO_2 networks. Adv Funct Mater 15:677

302. Ravirajan R, Bradley DDC, Nelson J, Haque SA, Durrant JR, Smit HJP, Kroon JM (2005) Efficient charge collection in hybrid polymer/TiO_2 solar cells using poly(ethylenedioxythiophene)/polystyrene sulphonate as hole collector. Appl Phys Lett 86:143101

303. Oey CC, Djurisic AB, Wang H, Man KKY, Chan WK, Xie MH, Leung YH, Pandey A, Nunzi J-M, Chui PC (2006) Polymer–TiO_2 solar cells: TiO_2 interconnected network for improved cell performance. Nanotechnology 17:706

304. van Hal PA, Wienk MM, Kroon JM, Verhees WJH, Slooff LH, van Gennip WJH, Jonkheijm P, Janssen RAJ (2003) Photoinduced electron transfer and photovoltaic response of a MDMO-PPV:TiO_2 bulk heterojunction. Adv Mater 15:118

305. Slooff LH, Kroon JM, Loos J, Koetse MM, Sweelssen J (2005) Influence of the relative humidity on the performance of polymer/TiO_2 photovoltaic cells. Adv Funct Mater 15:689

306. Feng W, Feng Y, Wu Z (2005) Ultrasonic-assisted synthesis of poly(3-hexylthio-phene)/TiO$_2$ nanocomposite and its photovoltaic characteristics. Jpn J Appl Phys 44:7494
307. Iijima S (1991) Helical microtubes of graphitic carbon. Nature 354:56
308. de Heer WA, Châtelain A, Ugarte D (1995) A carbon nanotube field-emission electron source. Science 270:1179
309. Romero DB, Carrard M, De Heer W, Zuppiroli L (1996) A carbon nanotube/organic semiconducting polymer heterojunction. Adv Mater 8:899
310. Curran SA, Ajayan PM, Blau WJ, Carroll DL, Coleman JN, Dalton AB, Davey AP, Drury A, McCarthy B, Maier S, Strevens A (1998) A composite from poly(m-phenylenevinylene-co-2,5-dioctoxy-p-phenylenevinylene) and carbon nanotubes: a novel material for molecular optoelectronics. Adv Mater 10:1091
311. Ago H, Petritsch K, Shaffer MSP, Windle AH, Friend RH (1999) Composites of carbon nanotubes and conjugated polymers for photovoltaic devices. Adv Mater 11:1281
312. Ago H, Kugler T, Cacialli F, Salaneck WR, Shaffer MSP, Windle AH, Friend RH (1999) Work functions and surface functional groups of multiwall carbon nanotubes. J Phys Chem B 103:8116
313. Ago H, Shaffer MSP, Ginger DS, Windle AH, Friend RH (2000) Electronic interaction between photoexcited poly(p-phenylene vinylene) and carbon nanotubes. Phys Rev B 61:2286
314. Kymakis E, Amaratunga GAJ (2002) Single-wall carbon nanotube/polymer photovoltaic devices. Appl Phys Lett 80:112
315. Kymakis E, Alexandrou I, Amaratunga GAJ (2003) High open-circuit voltage photovoltaic devices from carbon-nanotube–polymer composites. J Appl Phys 93:1764
316. Bhattacharyya S, Kymakis E, Amaratunga GAJ (2004) Photovoltaic properties of dye functionalized single-wall carbon nanotube/conjugated polymer devices. Chem Mater 16:4819
317. Kymakis E, Amaratunga GAJ (2005) Carbon nanotubes as electron acceptors in polymeric photovoltaics. Rev Adv Mater Sci 10:300
318. Kazaoui S, Minami N, Nalini B, Kim Y, Hara K (2005) Near-infrared photoconductive and photovoltaic devices using single-wall carbon nanotubes in conductive polymer films. J Appl Phys 98:084314
319. Rahman GMA, Guldi DM, Cagnoli R, Mucci A, Schenetti L, Vaccari L, Prato M (2005) Combining single-wall carbon nanotubes and photoactive polymers for photoconversion. J Am Chem Soc 127:10051
320. Guldi DM, Rahman GMA, Prato M, Jux N, Qin S, Ford W (2005) Single-wall carbon nanotubes as integrative building blocks for solar-energy conversion. Angew Chem Int Ed 44:2015
321. Landi BJ, Raffaelle RP, Castro SL, Bailey SG (2005) Single-wall carbon nanotube-polymer solar cells. Prog Photovoltaics Res Appl 13:165
322. Itoh E, Suzuki I, Miyairi K (2005) Field emission from carbon-nanotube-dispersed conducting polymer thin film and its application to photovoltaic devices. Jpn J Appl Phys 44:636
323. Rud JA, Lovell LS, Senn JW, Qiao Q, McLeskey JJT (2005) Water soluble polymer/carbon nanotube bulk heterojunction solar cells. J Mater Sci 40:1455
324. Kimura T, Ago H, Tobita M, Ohshima S, Kyotani M, Yumura M (2002) Polymer composites of carbon nanotubes aligned by a magnetic field. Adv Mater 14:1380
325. Cao J, Sun J-Z, Hong J, Li H-Y, Chen H-Z, Wang M (2004) Carbon nanotube/CdS core–shell nanowires prepared by a simple room-temperature chemical reduction method. Adv Mater 16:84

326. Robel I, Bunker BA, Kamat PV (2005) Single-walled carbon nanotube–CdS nanocomposites as light-harvesting assemblies: photoinduced charge-transfer interaction. Adv Mater 17:2458
327. Pradhan B, Batabyal SK, Pal AJ (2006) Functionalized carbon nanotubes in donor/acceptor-type photovoltaic devices. Appl Phys Lett 88:093106
328. Patyk RL, Lomba BS, Nogueira AF, Furtado CA, Santos AP, Mello RMQ, Micaroni L, Hümmelgen IA (2007) Carbon nanotube–polybithiophene photovoltaic devices with high open-circuit voltage. Phys Status Solidi RRL 1:43
329. Xu Z, Wu Y, Hu B, Ivanov IN, Geohegan DB (2005) Carbon nanotube effects on electroluminescence and photovoltaic response in conjugated polymers. Appl Phys Lett 87:263118
330. Pasquier AD, Unalan HE, Kanwal A, Miller S, Chhowalla M (2005) Conducting and transparent single-wall carbon nanotube electrodes for polymer–fullerene solar cells. Appl Phys Lett 87:203511
331. Ulbricht R, Jiang X, Lee S, Inoue K, Zhang M, Fang S, Baughman R, Zakhidov A (2006) Polymeric solar cell with oriented and strong transparent carbon nanotube anode. Phys Status Solidi B 243:3528
332. van de Lagemaat J, Barnes TM, Rumbles G, Shaheen SE, Coutts TJ, Weeks C, Levitsky I, Peltola J, Glatkowski P (2006) Organic solar cells with carbon nanotubes replacing In$_2$O$_3$:Sn as the transparent electrode. Appl Phys Lett 88:233503
333. Rowell MW, Topinka MA, McGehee MD, Prall H-J, Dennler G, Sariciftci NS, Hu L, Gruner G (2006) Organic solar cells with carbon nanotube network electrodes. Appl Phys Lett 88:233506

Adv Polym Sci (2008) 214: 87–147
DOI 10.1007/12_2007_119
© Springer-Verlag Berlin Heidelberg
Published online: 13 October 2007

Energy Harvesting in Synthetic Dendrimer Materials

Gemma D. D'Ambruoso · Dominic V. McGrath (✉)

Department of Chemistry, University of Arizona, Tucson, AZ 85721, USA
mcgrath@u.arizona.edu

Abstract In the past two and a half decades, dendrimers have emerged as a distinct branch of macromolecular chemistry. Tailoring of dendrimer structure yields precise placement of chromophores that can serve as energy harvesters, mimicking photosynthesis. The unique architecture afforded by dendrimers allows for multiple energy harvesters that can transfer their energy to a single core, which is important for optoelectronic applications such as organic light emitting diodes (OLEDs). This review emphasizes the energy transfer characteristics that these dendrimers provide rather then their synthesis.

Keywords Dendrimer · Light harvesting · Energy transfer · FRET

Abbreviations

ATP	adenosine triphosphate
BINOL	1,1′-bi-2-naphthol
C343	coumarin 343
C450	coumarin 450
CT	charge transfer
CZ	carbazole
2,3-dpp	2,3-*bis*(2-pyridyl)pyrazine
2,5-dpp	2,5-*bis*(2-pyridyl)pyrazine
DSB	distyrylbenzene
FRET	fluorescence resonance energy transfer
GPC	gel permeation chromatography
LC	ligand centered
2,3-Medpp$^+$	2-[2-(methylpyridiniumyl)]-3-(2-pyridyl)pyrazine
MLCT	metal-to-ligand charge transfer
NI	naphthalene imide
NIR	near infrared
NR	Nile Red
OLED	organic light-emitting diode
OPPV	oligo (*p*-phenylene vinylene)
OXZ	oxadiazole
PA	phenylacetylene
PDI	perylene diimide
PET	photo-induced electron transfer
P_{FB}	free base porphyrin
PI	perylene imide
PL	photoluminescence
PPI	poly(propylene imine)
PPV	poly(*p*-phenylene vinylene)
P_{Zn}	zinc-metallated porphyrin
TDI	terrylene diimide
TI	terryleneimide
TPA	two-photon absorbing
TPE	two-photon excitation

1
Introduction

Dendrimers are an intriguing scaffold for constructing energy harvesting systems since their architectures roughly mimic that of natural photosynthetic centers in which light is harvested by chlorophyll chromophores encircling a reactive core. Energy transfer from the outer chlorophyll chromophores to the reaction center ultimately results in ATP production, which is essential for life. This chapter reviews artificial energy-harvesting systems that employ a dendritic architecture.

The mechanism of energy transfer observed in dendrimers is typically fluorescence (or Förster) resonance energy transfer (FRET). FRET is the radiationless transfer of excitation energy from a "donor" chromophore to an appropriately positioned "acceptor" chromophore through long-range dipole-dipole coupling, with the return of the donor to the ground state (Fig. 1). FRET efficiency is determined by the distance between the donor and acceptor (typically ~1–10 nm), the spectral overlap between the donor emission spectrum and the acceptor absorbance spectrum, and the relative orientation of the donor and acceptor dipole moments.

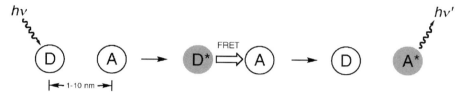

Fig. 1 Schematic of the FRET process. Excitation of the fluorescent donor is followed by radiationless transfer of energy to the acceptor chromophore, returning the donor to the ground state. FRET is then usually, although not always, manifested by subsequent fluorescence of the acceptor

The role of dendrimers in light-harvesting systems varies substantially. First, the dendrimers can act as an inert spacer to separate two different chromophores (host/guest) from each other. In these cases, the dendrimers do not participate in the energy transfer between the chromophores, as their excited state energy is typically higher than both the host and guest. In addition, increasing dendrimers generation increases host/guest distance and chromophore communication can begin to decrease (as above the Förster radius) as a result. Second, the dendrimers subunits can act as the host chromophores and are responsible for light harvesting. In this case, light absorbed by the dendrimer (acting as the host) is transferred to a core moiety (guest) from which excited state photophysical processes, such as emission or electron transfer may ensue. Increasing the generation number often allows for greater light harvesting capabilities, since the number of dendrimers subunits increases.

In both cases (dendrimers as spacers or dendrimers as hosts), dendrimers offer a unique advantage over classical energy transfer between a chromophoric pair in that the number of host chromophores can far exceed the number of guest chromophores, and is a strict function of generation. The effective molar absorptivity of the host chromophores can therefore be increased (predictably) and the dendrimer itself can be a better light harvester then a single host chromophore. Better light harvesting and an increase in the amount of energy transferred to the guest results in a greater amount of emission from the dendrimers and is referred to as the "antennae effect."

In addition to the antennae effect, dendrimers are also effective insulators and exhibit the "shell effect." In providing a dense shell around the incorporated chromophores, dendrimers effectively prevent aggregation which leads to non-emissive excimers and self-quenching that occurs when chromophores with small Stokes shifts are within short distances of one another. This "shell effect" allows for increased photoluminescence efficiency of the enclosed chromophore, which is important for optoelectronic devices.

We have organized this review by type of chromophore and/or dendrimer structure.

2
Metal-containing Dendrimers

Light-harvesting metal containing dendrimers exist in two categories: (i) dendrimers where metals are at both the core and branching points [1–9] and (ii) dendrimers where metal cores are surrounded by aromatic dendrons [10–19]. In both cases, light-harvesting followed by energy transfer to the metal occurs. These two categories are addressed here.

2.1
Metal-containing Dendrons and Cores

In the early 1990s, Balzani and coworkers began to synthesize decanuclear homo- and heterometallic dendrimers that contain Ru(II) and Os(II) surrounded by nitrogen-containing ligands (bpy derivatives). The metals occupied sites in both the dendrons and at the core of the dendrimers [1, 6–9, 20]. Although the Balzani group had for some time been synthesizing multimetallic species in tri- [21], tetra- [22–24], hexa- [25], and heptanuclear [26] complexes, this is the first work that displays these complexes in a dendritic fashion. The dendrimers were synthesized in a "complexes as ligands/complexes as metals" manner in which metal complexes (instead of individual metals and ligands) are used as the reactive species. This approach has produced metallodendrimers in which the number of metals per dendrimer is as high as 22 [2, 3, 5]. The dendrimers consist of bridging ligands 2,3- and 2,5-*bis*(2-pyridyl)-pyrazine (2,3-dpp **2** and 2,5-dpp **3**) that connect the metal centers, and terminal ligands, 2,2′-bipyridine (bpy) **1**, 2,2′-bisquinoline (biq) **5** and 2-[2-(methylpyridiniumyl)]-3-(2-pyridyl) pyrazine (2,3-Medpp$^+$) **4** that "cap" the peripheral metals (Scheme 1).

Light harvesting in these systems arises from overlap of the emission band associated with ligand-centered (LC) transitions in the UV as well and several metal-to-ligand charge-transfer (MLCT) transitions in the visible region. Each metal center in the dendrimer is capable of absorbing light, so the molar absorptivity and amount of harvested light increases with increasing number

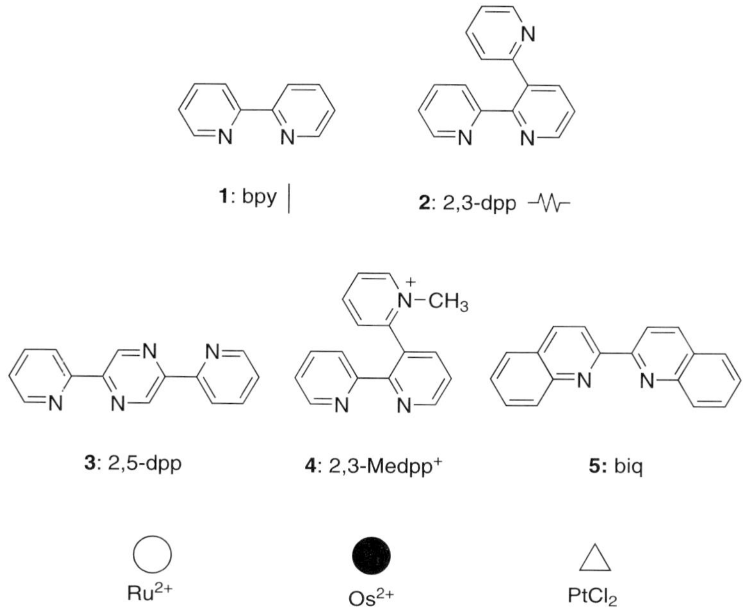

1: bpy

2: 2,3-dpp

3: 2,5-dpp

4: 2,3-Medpp$^+$

5: biq

Ru^{2+}

Os^{2+}

PtCl$_2$

Scheme 1 Building blocks for multimetallic light harvesting dendrimers [4, 5]

of metal sites. Emission from these dendrimers is primarily through radiative decay of the lowest lying metal-to-ligand charge-transfer triplet state, ^{3}MLCT, for both Ru(II) and Os(II) cases [27]. In the case of the homometallic Ru(II) dendrimers, emission occurs solely from the lower-energy peripheral Ru(II) complexes which are coordinated to the terminal ligands (either **1** or **5**). In the Ru(II)/Os(II) mixed-metal systems, emission is expected solely from the lower energy Os cores (Scheme 2). However, in these mixed metal den-

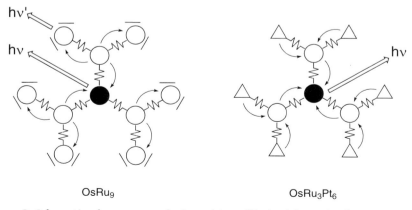

OsRu$_9$

OsRu$_3$Pt$_6$

Scheme 2 Schematic of energy transfer in multimetallic dendrimers [4, 5]

drimers, replacing the Ru(II) at any of the metal sites with Os(II) results in only partial energy transfer from the Ru(II) sites (both interior and peripheral) to the Os(II) units, and therefore emission from both.

Incomplete energy transfer from the Ru(II) sites to the Os(II) cores are thought to be due to the "blocking" ability of the high-energy interior Ru(II) units. The energies of the lowest excited states follow in the order of Ru(II) (interior) > Ru(II) (exterior) > Os(II) (center), and therefore light harvested at the Ru(II) periphery (or transferred there from the interior) cannot progress completely to the center Os(II) by energy transfer due to the blocking interior Ru(II) units (Scheme 2) [1]. Hence, in order to create a true antennae species in which light harvested is transferred efficiently to the core, decanuclear dendrimers were prepared in which the peripheral Ru(II) units are replaced by Pt(II) whose lowest excited state resides above that of the interior Ru units [4]. Light harvested by an $OsRu_3Pt_6$ dendrimer gives rise to

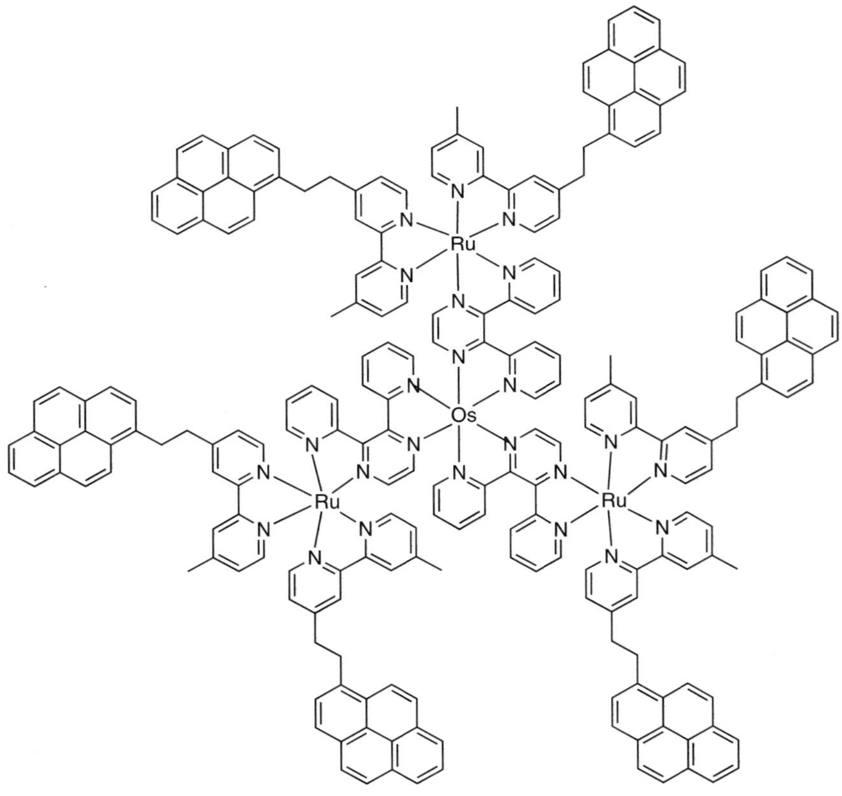

6

Scheme 3 Structure of 6 from Campagna and coworkers [28]

a sole emission band at 875 nm which is due to the emitting ^{3}MLCT of the Os core unit. These dendrimers represent the first metallodendrimers containing three different types of metals.

To introduce organic chromophores into multinuclear metallo-organic dendrimers, Campagna and coworkers have placed pyrene moieties on the periphery of a Ru(II)- and Os(II)-containing dendrimer 6 (Scheme 3) [28]. An enhanced antennae effect occurs in the dendrimer because of intramolecular pyrene-to-ligand CT transitions, which are seen in the visible region of the absorption spectra, in addition to the absorption bands due to the individual components of the dendrimer. The energies of the lowest excited-state are in the order: pyrene > Ru(II) interior > Os(II) core, and therefore energy is funneled quantitatively to the Os(II) core from where emission is seen around 800 nm.

2.2
Metal-cored Dendrimers with Organic Dendrons

The ability to alter the photo- and electrochemical properties of a reactive species by surrounding the unit with an organic dendrimer framework has been studied extensively [29]. In the case of metals, aromatic (bulky) dendrons have been used to insulate the metals from dioxygen and other excited-state quenchers [10, 19, 30, 31], prevent aggregation [17–19], and in most cases, harvest light in the higher-energy dendrons which can be transferred to the lower-energy metal core [10–12, 17, 19, 32].

2.2.1
Ru(bipy)$_3$ Cores

Balzani and coworkers have synthesized first and second generation Ru(II) cored dendrimers with napthyl groups (12 or 24) at the periphery of benzyl(aryl ether) dendrons [10]. Dendrons 7 and 8 were attached to bipyridine ligands (bpy) and complexed with Ru(II) to give the resulting dendrimers 9 and 10 (Scheme 4).

Photoexcitation of the napthyl and dialkoxybenzyl groups of the dendrons 7 and 8 give rise to an emission band near 600 nm, which is attributed to the [Ru(bpy)$_3$]$^{2+}$ core complex (Fig. 1) [27]. The emission of the napthyl groups at 330 nm is nearly quenched, indicating an energy transfer from the peripheral napthyl groups to the [Ru(bpy)$_3$]$^{2+}$ core. Longer luminescence lifetimes of the dendrimers in aerated solutions (compared to the [Ru(bpy)$_3$]$^{2+}$ parent) indicate that the bulky dendrons prevent dioxygen from accessing the core and quenching the excited state, which has been shown for other Ru(II) containing dendrimers [30, 31].

A similar study by Castellano and coworkers placed coumarin 450 (C450) dye molecules on the periphery of "reverse" benzyl aryl ether dendrons, first

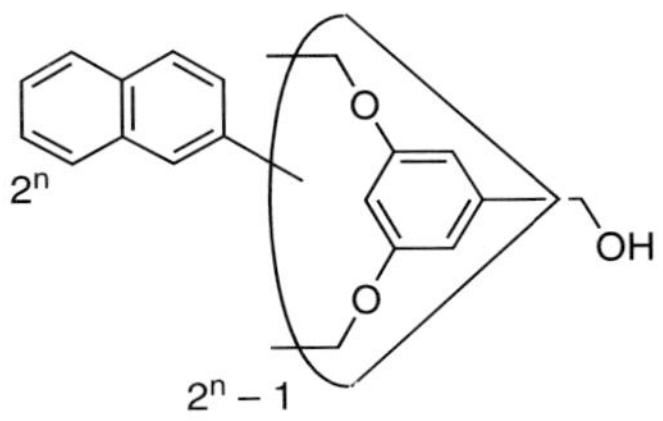

7: n = 1, [G1]
8: n = 2, [G2]

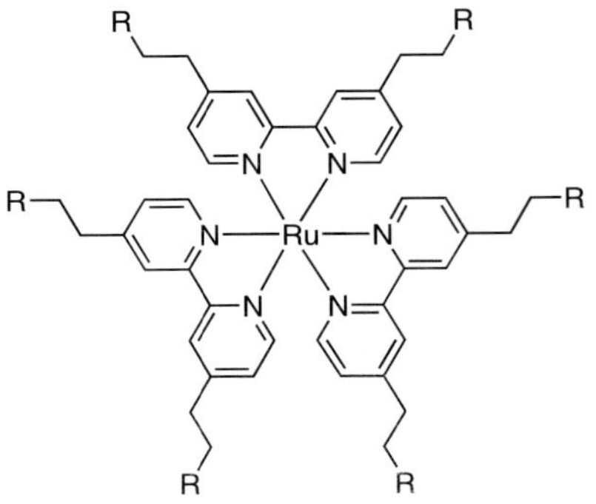

9: R = [G1]
10: R = [G2]

Scheme 4 Structures of **7–10** from Balzani and coworkers [10]

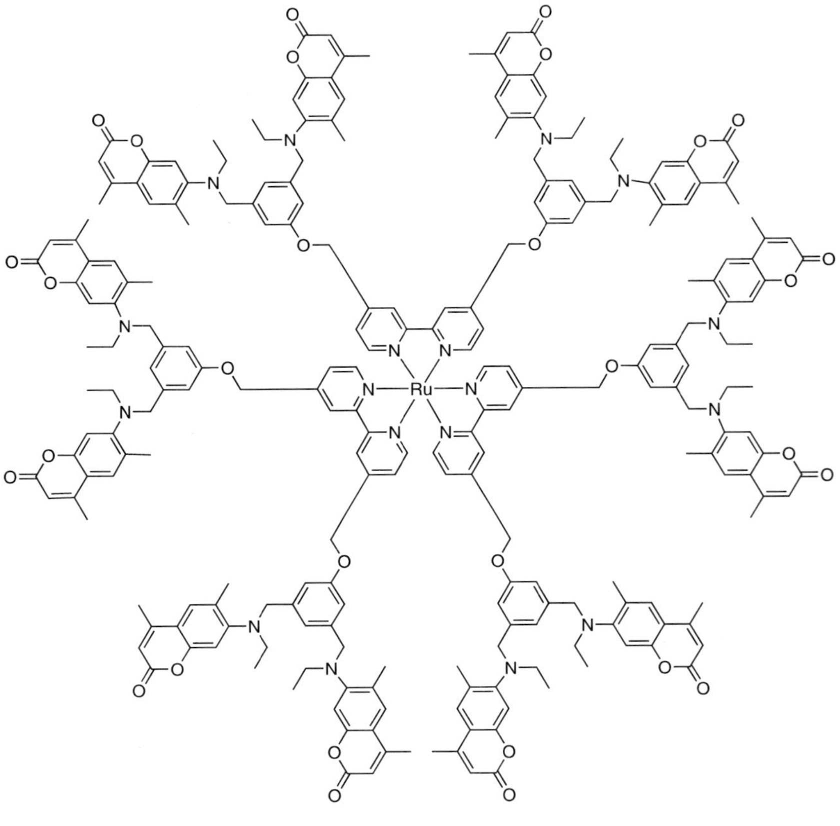

11

Scheme 5 Structure of **11** from Castellano and coworkers [11]

introduced by Hanson [33], attached to a Ru(II) core moiety to give dendrimer **11** (Scheme 5) [11]. The chromophores satisfy the requirements of a Förster energy transfer as the emission of the C450 significantly overlaps the absorption of the $[Ru(bpy)_3]^{2+}$ species and give a calculated Förster radius of 41.2 Å. Fluorescence spectra indicate that light absorbed by the C450 moieties on the periphery of the first generation dendrimer **11** is accompanied by energy transfer and subsequent emission by the $[Ru(bpy)_3]^{2+}$ species at the core (Fig. 2). Estimated energy transfer efficiency was calculated by comparing the absorbance and excitation spectra (observed at emission wavelength of $[Ru(bpy)_3]^{2+}$ complex, 610 nm) and was determined to be around $95 \pm 10\%$. Preliminary excited state lifetime measurements indicated that dioxygen quenching was reduced by the presence of the "shielding" dendrons [11]. However, further investigation of the bimolecular quenching rates for several additional quenchers indicated that dioxygen was an exception rather than a rule [32]. In all other cases the dendrons allowed access to the core by

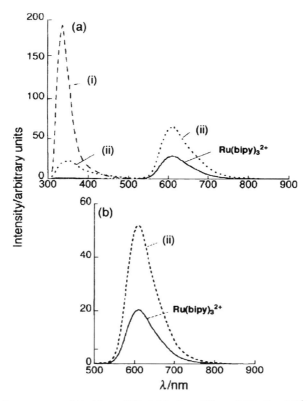

Fig. 2 a Emission spectra of 10 (*dots*, (ii)), 8 (*dashes*, (i)), and $[Ru(bpy)3]^{2+}$ (*solid*) when exciting into napthyl groups (270 nm). **b** Emission of 10 (*dots*, (ii)) and $[Ru(bpy)3]^{2+}$ when exciting into the $[Ru(bpy)3]^{2+}$ core (450 nm)

the quenchers (e.g. 9-methylanthracene and phenothiazine). Therefore, the reduced quenching in the dioxygen case was possibly due to the lower solubility of dioxygen in the microenvironment surrounding the metal core, not an inability to diffuse to the core.

Dehaen and coworkers utilized carbazole dendrons attached to 1,10-phenanthroline ligands (phen) to construct a dendrimer around a Ru(II) core (12, Scheme 6) [12]. Absorbance transitions assigned to carbazole, phen, and carbazole-to-phen CT transitions all gave rise to emission from the Ru(II) core, providing yet another light-harvesting dendrimer, which illustrates the antennae effect.

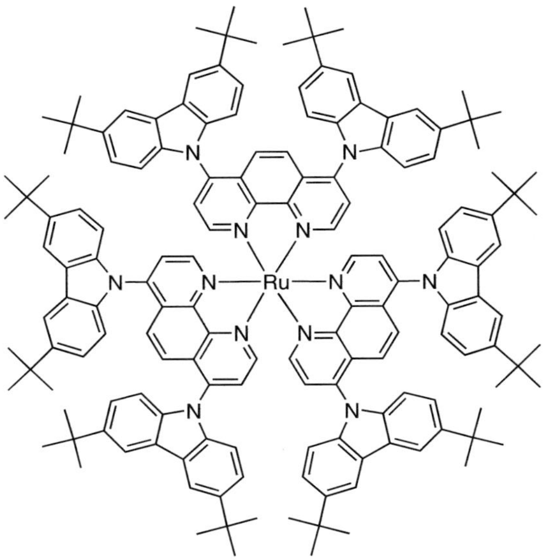

12

Scheme 6 Structure of **12** from Dehaen and coworkers [12]

Further examples of harvesting dendrimers that contain transition metals include poly(propylene imine) (PPI) dendrimers modified with 32 dansyl groups at the periphery (**13**, Scheme 7) [13], whose fluorescence is quenched when a single Co^{2+} ion is coordinated to the interior amines, thereby creating a fluorescent chemosensor (Fig. 3) [34, 35]. The quenching arises from energy and electron transfer between the excited dansyl groups and the Co(II) amine complexes. Similar results have been obtained for lanthanide metals coordinated to the interior of polylysine dendrimers containing dansyl groups on the periphery [14–16].

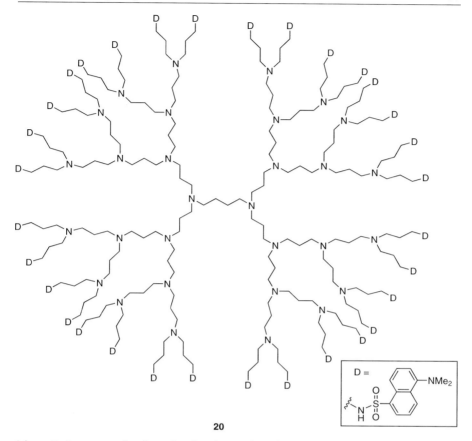

20

Scheme 7 Structure of **13** by Balzani and coworkers [13]

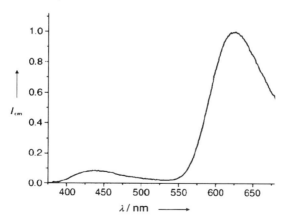

Fig. 3 Emission spectrum of dendrimer **11** in CH₃CN when excited into the coumarin 450 dyes (343 nm) [11]

2.2.2
Lanthanide Cores

Because of their narrow emission bands, long luminescence lifetimes, and absorption profiles that span from the near-UV to the near-IR, lanthanides have been investigated for applications such as light-emitting diodes [36] and optical amplifiers [37]. However, the tendency of lanthanides to aggregate leads to self-quenching, which limits their luminescence efficiency and lifetimes. Also, proximity to solvent molecules which can act as quenchers limits their potential use as optoelectronic materials [38, 39].

In work in lanthanide containing dendrimers, Kawa and Fréchet [17, 18] synthesized benzyl(aryl ether) dendrons with carboxylate focal points that formed 3 : 1 complexes with lanthanide ions Er^{3+}, Tb^{3+}, or Eu^{3+} (Scheme 8). Emission spectra for **14b**, **15b**, **16b** at identical concentrations show an increase in the emission from the Eu core when the dendrons are excited. This is due to both an increase in energy harvesters as generation increases (antennae effect) and an increase in site-isolation allowing for reduced self-quenching of the Eu cores (shell effect). Interestingly, when the substitution of the phenyl ring closest to the focal point is changed from 3,5- to 2,5-, emission enhancement is not as pronounced.

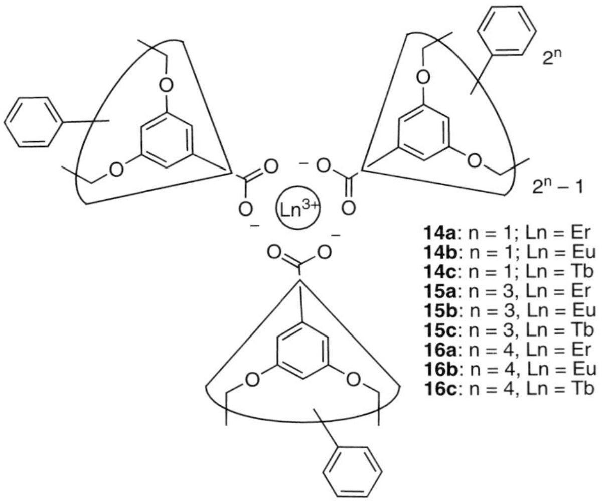

14a: n = 1; Ln = Er
14b: n = 1; Ln = Eu
14c: n = 1; Ln = Tb
15a: n = 3, Ln = Er
15b: n = 3, Ln = Eu
15c: n = 3, Ln = Tb
16a: n = 4, Ln = Er
16b: n = 4, Ln = Eu
16c: n = 4, Ln = Tb

Scheme 8 Structure of lanthanide-cored dendrimers of Kawa and Fréchet [17, 18]

Similar Er^{3+}, Tb^{3+}, and Eu^{3+} dendrimers constructed from dendrons with a benzoic acid ligating moiety but aliphatic polyester scaffold of several generations and fluorinated phenyl groups on the periphery were recently reported (Scheme 9) [19]. These structures were designed to minimize moisture

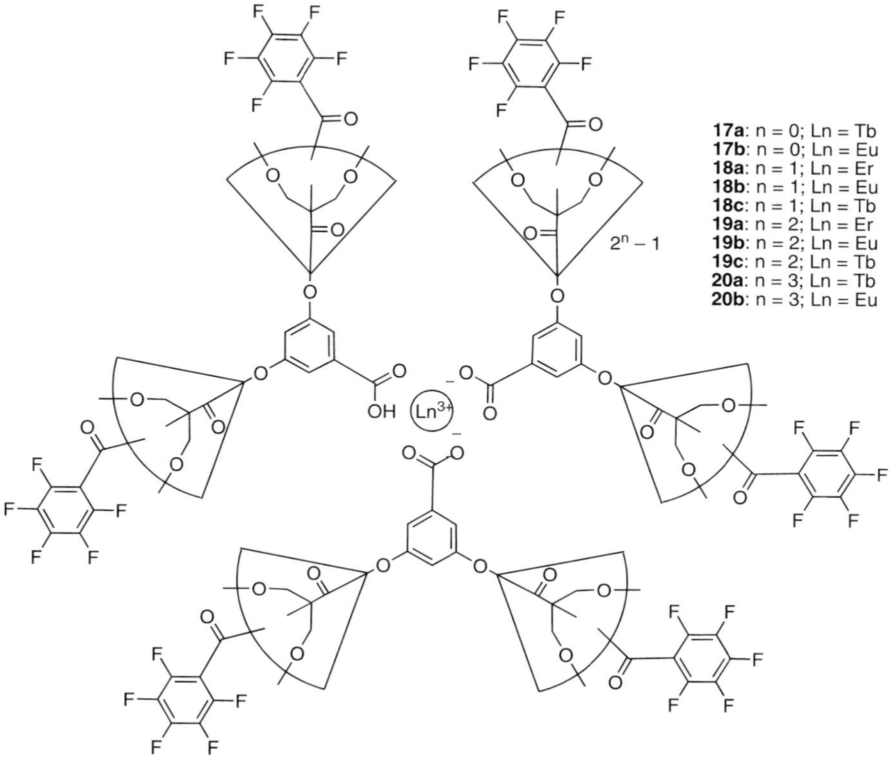

17a: n = 0; Ln = Tb
17b: n = 0; Ln = Eu
18a: n = 1; Ln = Er
18b: n = 1; Ln = Eu
18c: n = 1; Ln = Tb
19a: n = 2; Ln = Er
19b: n = 2; Ln = Eu
19c: n = 2; Ln = Tb
20a: n = 3; Ln = Tb
20b: n = 3; Ln = Eu

Scheme 9 Structure of lanthanide-cored dendrimers of Hult and coworkers [19]

penetration into the core unit and reduce optical absorption in the near-IR region. Energy transfer from the perfluorinated phenyl peripheral units to the Eu and Tb cores were studied spectrophotometrically in solution. Excitation of the perfluorinated phenyl groups decays in ca. 0.7 ns, while complexation to the Ln^{3+} core results in a longer decay time (10–13 ns) in the nanosecond range. Lanthanide luminescence decays in the 1-1.5 ms range. Excitation of the peripheral phenyl groups was transferred to the lanthanide core, and the main quenching mechanism appeared to be vibrational loss to surrounding C–H bonds of the dendritic shells.

3
Phenylacetylene Dendrimers

Dendrimers consisting of phenylacetylene moieties exist as (i) "compact" or "extended" meta-conjugated phenylacetylene (PA) dendrimers [40–54], and (ii) unsymmetrical branched phenylacetylene dendrimers [55–58].

3.1
"Compact" and "extended" Phenylacetylene Dendrimers

Moore and coworkers have synthesized phenylacetylene dendrimers of two types, "compact" (e.g. **21**) and "extended" (e.g. **22**) [40–42]. Both the "compact" and "extended" forms contain meta-substituted phenylacetylene subunits. However, the lengths of the phenylacetylene segments differ between the two types. In the "compact" form, all the conjugated segments are effectively the same length and exhibit the same excitation energies; meta-substitution ensures that π-conjugation is blocked at each branching point, allowing for a localized excited state upon excitation [43, 46]. In the "extended" form, as the dendrimer proceeds from the periphery to the core, the conjugation length of each generation increases by one phenylacetylene moiety. Concomitant with the increase in π-conjugation is a step-wise decrease in the lowest excited state energy. This allows the dendrimer to act as an energy "funnel". The "extended" dendrimer still contains localized excitations. However, in each layer leading up to the core, the excited state energy is more delocalized due to increasing conjugation. It is important to emphasize that neither type of dendrimer exhibits a delocalized excitation over the entire dendrimer. Rather, energy is localized in each layer or generation.

Tris dendrimers (three dendrons attached to a 1,3,5-trisubstituted phenyl core) in "compact" [40, 42] and "extended" [41] forms have been synthesized as well as "nanostars" [44, 45] (Scheme 10) which consist of a "compact" or "extended" dendron attached to a perylene "trap" (26 and 27). Their energy transfer rates and mechanisms have been studied extensively [43, 45–52] Important features are observed when comparing the absorption spectra for a series of "compact" and "extended" dendrimers (Fig. 4) [43, 48]. In the "compact" case, the absorption maxima remain at a fixed wavelength throughout the series with only the intensity scaling with increasing number of absorbing chromophores. This indicates that not only are the excitations localized within the layers, but also that the chromophores, regardless of location within the dendrimer, all absorb at the exact same wavelength. Time-resolved fluorescence and fluorescence anisotropy dynamics of "compact" structures, however, have revealed delocalized excited states [59, 60].

The spectra of the "extended" series, however, contain additional peaks of lower energy with each generation, corresponding to the linear segments of extended conjugation. A broader absorption spectra for "extended" dendrimers imply that these molecules will be more effective light harvesters. It should be noted that the absorption spectra of the "compact" and "extended" nanostars are similar to their respective dendrimers with an additional peak at 475 nm for perylene absorption.

The quantum yield for energy transfer Φ_{ENT} for "compact" and "extended" nanostars was measured by comparing absorption and excitation spectra normalized in the perylene absorption region (430–500 nm) (Fig. 5) [45, 47].

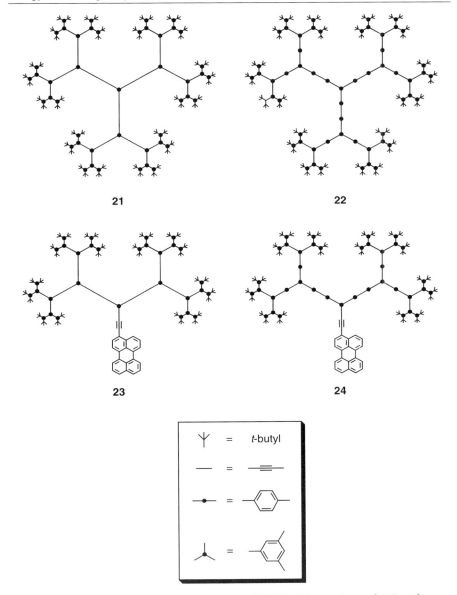

Scheme 10 Structures of "compact" and "extended" dendrimers (**21** and **22**) and nano-stars (**23** and **24**) [40–42, 44, 45]

For the "extended" nanostar **24**, Φ_{ENT} was 98%. The "compact" nanostars of varying generations give Φ_{ENT} of 95% for the lowest generation and 54% for the highest; residual emission from the phenylacetylene segments increases with generation. Fluorescence lifetime measurements also differ greatly for

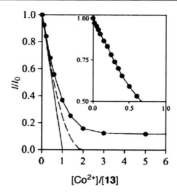

Fig. 4 Fluorescence intensity of **13** with the addition of Co^{2+} (*dark circles*). *Inset* shows linearity at low concentration. *Solid line* is linear extension of the initial slope, *dashed line* is expected if two Co^{2+} are independently coordinated in **13** [35]

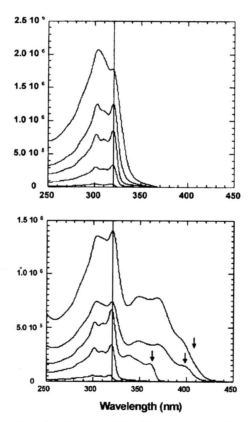

Fig. 5 Absorption spectrum for "compact" dendrimers (*top*) and "extended" dendrimers (*bottom*) in hexane. *Arrows* indicate the absorption bands for two, three and four phenyls in a phenylacetylene chain, respectively [48]

the "extended" (270 ps) and "compact" (2.3 ns) structures, yielding rate constants for energy transfer k_{ET} for the "compact" molecules that are lower by two orders of magnitude than those for the "extended" structure. Although the Förster equation provides k_{ET} values that cannot be correlated to the observed ones, the reasons for lower efficiency in the "compact" structures can be explained by the Förster model: as generation increases, chromophore separation becomes too great for efficient energy transfer. The energy "funnel" inherent in "extended" structures, on the other hand, allows for efficient stepwise transfer of energy to the perylene core, which is not hampered by the interchromophoric distance as in the "compact" case.

Although at higher generations "compact" structures have decreased Φ_{ENT} values, direct sensitization of the perylene cores by the diphenylacetylene segments in the dendrons can occur. In the "extended" nanostar **24**, perylene excimer formation lends evidence that long-range energy transfer also takes place, in addition to a multistep process [48, 49]. Although a concentration excimer dependence was expected and observed, an unexpected dependence on the wavelength of excitation was also observed, as measured by steady-state fluorescence. Excimer fluorescence was highest when the innermost layer of phenylacetylene moieties was excited, indicating that a phenylacetylene excimer is initially formed which in turn excites a perylene excimer (Table 1). Because the excimer emission was much lower when interior or periphery phenylacetylenes were excited, all excitons created in these outer regions do not always funnel to the interior ones, leaving the possibility that there is some amount of direct energy transfer that does take place between the periphery and the perylene trap.

Pu and coworkers are interested in using chiral dendrimers as enantioselective fluorosensors [53, 54]. By utilizing the antennae effect available to light-harvesting dendrimers, fluorescence amplification can be achieved, implying that the dendrimers will be more sensitive fluorosensors to smaller amounts of a quencher. The dendrimer **25** consists of "compact" phenylacetylene dendrons surrounding a chiral 1,1'-bi-2-naphthol (BINOL) core (Scheme 11). Precise placement of the dendrons around the BINOL core creates a chromophore that has extended conjugation and therefore acts as an

Table 1 Excimer/monomer ratio for "extended" nanostar **27** [48, 49]

Exitation wavelength, nm	Dendrimer generation	Excimer/ monomer ratio
310	3	0.33
334	2	0.31
390	1	4.0
450	Perylene trap	0.14

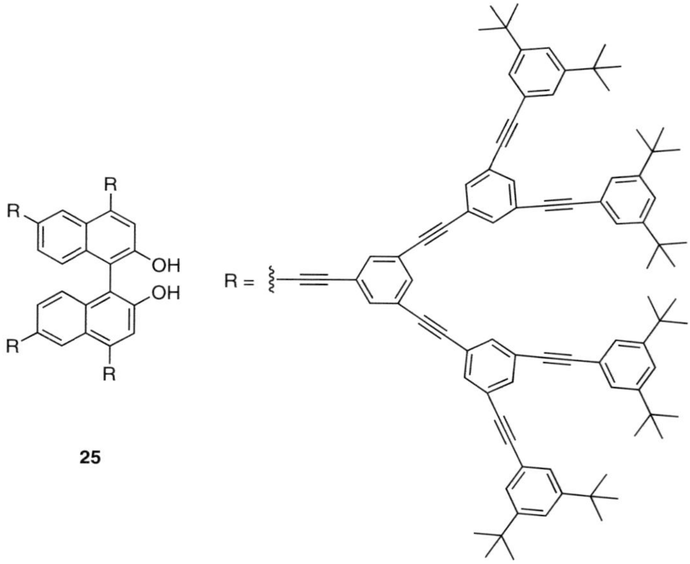

Scheme 11 Structure of **25** by Pu and coworkers [53, 54]

energy sink for energy harvested by the dendrons. Hydrogen bonds formed between the BINOL core and chiral amino alcohols in solution quench the fluorescence from the dendron/core chromophore. Stern–Völmer plots show a mirror-image relationship between the enantiomers of the dendrimers and the enantiomers of the quenchers (e.g. (S)-dendrimer quenches (S)-amino alcohol more efficiently and (R)-dendrimers quench (R)-amino alcohols more efficiently).

3.2
Unsymmetrical Phenylacetylene Dendrimers

If *meta* and *para* branching patterns are utilized, instead of the all-*meta* branching in Moore's PA dendrimers, dendrons now consist of a long linear segment that is attached to the core and from which PA segments of different length are attached (**26a–d**) as reported by Peng (Scheme 12) [55–58]. The overall effect is one of an energy "short-cut" as now all segments varying in length are directly attached to the lowest energy PA moiety. In other words, the number of steps in this multistep energy transfer is diminished from Moore's "extended" system. Energy transfer as measured by Φ_{ENT} is very efficient (85–95%) and does not seem to decrease with increasing generation as in the "compact" symmetrical PA dendrimers. Also interesting is the minor effect that dendron geometry has on the Φ_{ENT}. Dendrons were attached to perylene "traps" in two ways; the perylene was either conjugated

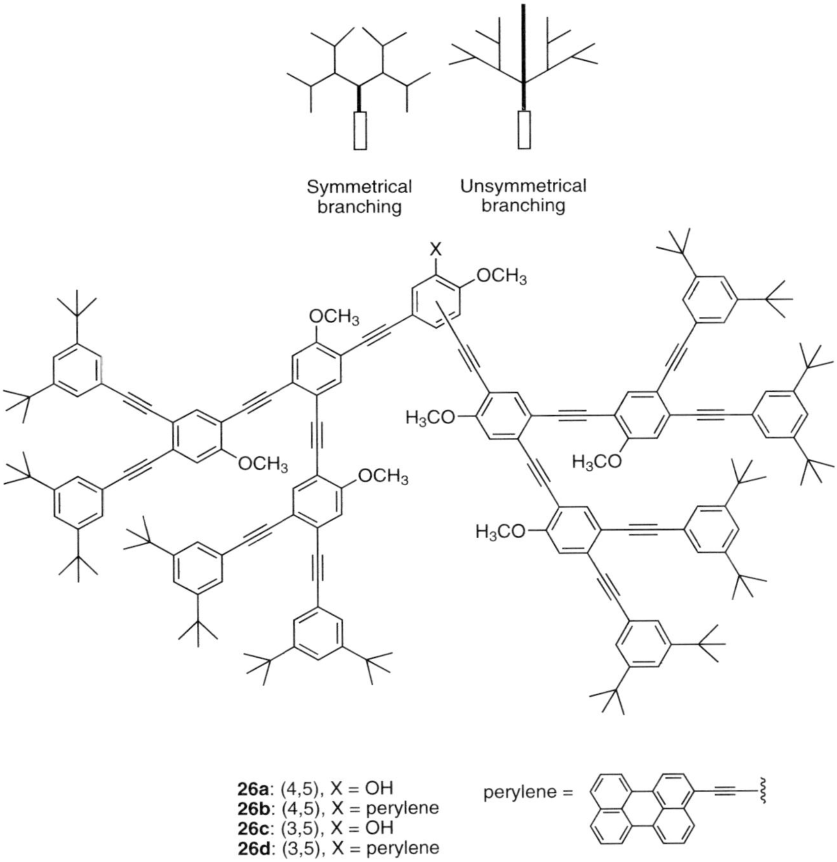

26a: (4,5), X = OH
26b: (4,5), X = perylene
26c: (3,5), X = OH
26d: (3,5), X = perylene

perylene =

Scheme 12 Structures of unsymmetrical phenylacetylene dendrimers **26a–d** by Peng and coworkers [55–58]

to the longest linear segment of the dendrons or attached by a *meta*-linkage, disrupting the conjugation. Only slightly lower Φ_{ENT} were obtained for the *meta*-linked [G1] dendrimers (difference of 3%) but the difference increased to 9% at the [G2] level.

4
Dendrimers Containing Distyrylbenzene or Stilbene Units

Although early examples of OLEDs included polymers, specifically poly (*p*-phenylene vinylene) (PPV), as the emissive material [61], conjugated oligomers can also be useful emitters as well as energy harvesters. Oligomers offer advantages over their polymeric parents such as ease of synthesis and functionalization allowing for easy introduction of various substituents. They can also be incorporated into dendrimers as both host and guest chromophores.

Distyrylbenzenes (DSB), also known as oligoPPVs, have been extensively functionalized to alter their band-gap properties as well as their two-photon cross-sections. By tailoring the end-capping styrene groups to be branching units in dendrons, several dendrimers have been synthesized containing DSB at the core. Kwok and Wong [62] have placed either one or two poly (aryl ether) dendrons around a core DSB to elucidate the influence of the dendritic wedge on aggregation of the cores (Scheme 13). By placing a single electron-transporting oxadiazole on the exterior of the dendrons to offset the hole-affinity of the DSB, the dendrimers (28 and 30) then become more charge-balanced.

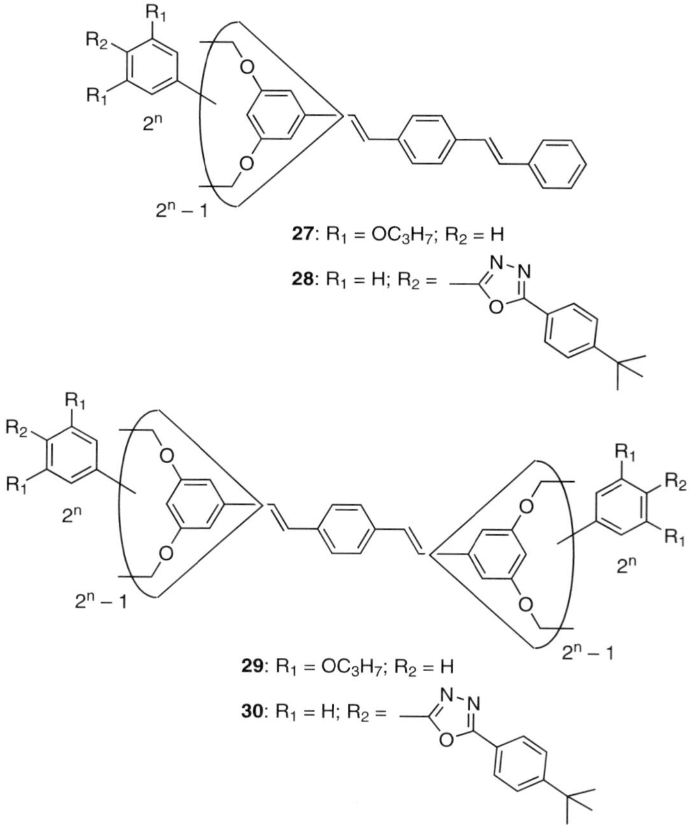

Scheme 13 Distyrylbenzene dendrimers 27–30 by Kwok and Wong [62]

Fluorescence lifetimes and quantum yields of the symmetrical dendrons (27 and 28) were higher then those of their asymmetric counterparts (29 and 30) indicating that there is more effective shielding of the DSB cores by two dendrons than one. In all other areas, such as energy transfer from the dendrons to the DSB core and OLED device performance, asymmetrical dendrimers seem to hold the advantage over the symmetrical ones. Performance is further enhanced in devices when a hole-transporting material, diphenylamine, is blended into the emissive layer of the oxadiazole asymmetrical dendrons, indicating that placing an oxadiazole on every dendron is excessive.

Samuel and Burn [63] have used DSB cores surrounded by conjugated dendrons as the emissive and charge transport layer in OLEDs. The conjugated dendrons consist of stilbene units in a meta-branching pattern, which prevents uninterrupted conjugation throughout the dendron (Scheme 14). Each stilbene subunit therefore appears electronically isolated, although red-shifted absorbance and PL spectra indicate that there is some amount of delocalization. Excitation of the stilbene units does result in emission solely from the distyrylbenzene core in both solution and thin films, giving evidence of efficient energy transfer. Light-emitting devices from all three gener-

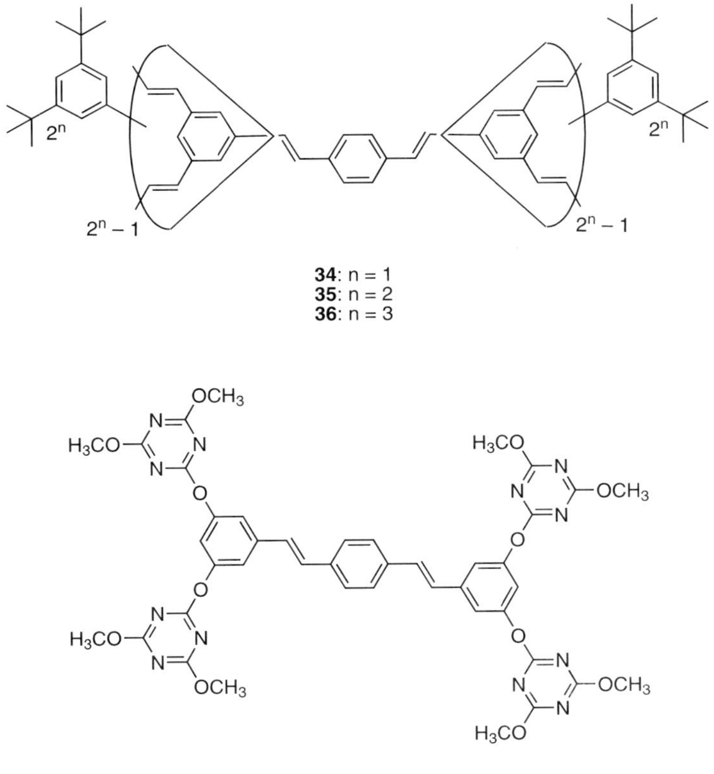

34: n = 1
35: n = 2
36: n = 3

37

Scheme 14 Distyrylbenzene cored dendrimers **31–34** by Samuel and Burn [63, 64]

ations of dendrimers were equal to (**31** case) or better than (**32** and **33** cases) single-layer devices containing PPV. Dendrimers with DSB cores surrounded by electron transporting triazine groups (**34**) were found to produce comparable PL quantum yields, but lower device external quantum efficiencies by a factor of 10 [64].

Burn and Samuel have used their stilbene dendrons to transfer energy to a number of different cores for color-tuning in OLEDs (Scheme 15) [65]. First and second generation dendrons with an aldehyde at the focal point have been coupled with the bisphosphonate esters to form DSB and distyrylanthracene cored dendrimers **35–40**. Dendron aldehydes and pyrroles were condensed to form porphyrin dendrimers **41** and **42**. Proton NMR of the porphyrin cored dendrimers **41** and **42** showed a chemical shift of the *tert*-butyl groups on the periphery of the dendrons, indicating that the *tert*-butyl groups were actually hovering over the porphyrins. This points to an orthogonal

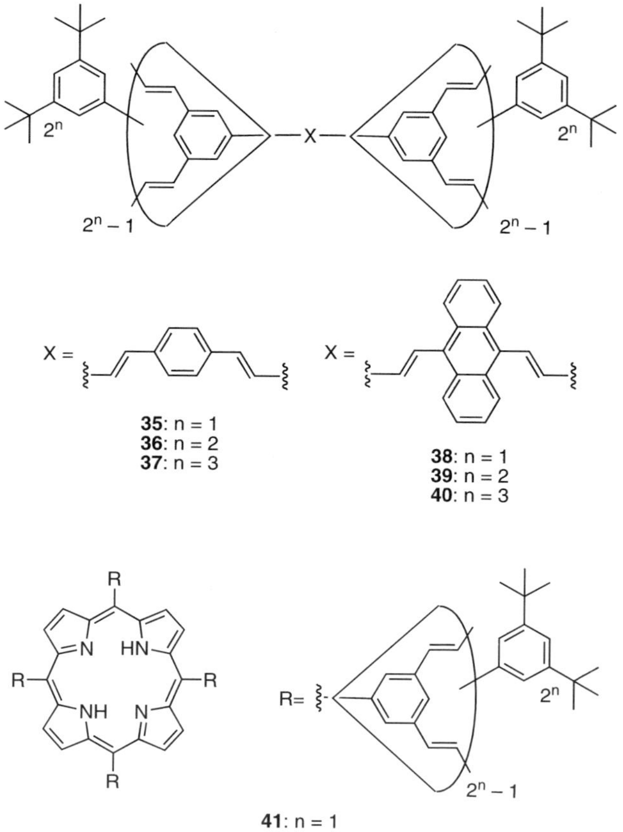

35: n = 1
36: n = 2
37: n = 3

38: n = 1
39: n = 2
40: n = 3

41: n = 1
42: n = 2

Scheme 15 Stilbene dendrons by Burn and Samuel with DSB (**35–37**), distyrylanthracene (**38–40**), and porphyrin cores (**41** and **44**) [65]

positioning of the dendrons to the porphyrin core, and the authors were interested as to whether this would hinder energy transfer from the dendrons to the porphyrin core. Interestingly, in dendrimers **35-40**, energy transfer was found to be around 65%. In the case of the porphyrin dendrimers **41** and **42**, however, excitation of the dendrons resulted in quantitative energy transfer to the core as measured by comparing excitation and absorbance spectra.

5
Porphyrin Containing Dendrimers

The three categories of porphyrin-containing dendrimers that will be discussed include: (i) porphyrin cores surrounded by organic dendrons [66–69], (ii) dendritic arrays of porphyrins at the both core and branching points [70–72], (iii) organic dendrimers with porphyrins at the periphery [73–75], and (iv) porphyrin cores with a donor chromophore periphery [76–79].

5.1
Dendrons Surrounding a Porphyrin Core

Work by Jiang and Aida on light-harvesting and energy transfer properties of porphyrin-containing dendrimers began as an extension of their investigation of azobenzene-containing dendrimers and their energy transfer phenomena (Scheme 16) [80–82]. Mono-azobenzene-containing dendrimers **43a–d** exhibited remarkable generation dependent energy harvesting capabilities. As expected, compounds **43a–d** undergo ultraviolet photoinduced isomerization from the E-form to the Z-form of the core azobenzene and thermally isomerize in the reverse direction with typical half-lives. However, when individual solutions of the Z-forms of **43c** and **43d** were irradiated with infrared irradiation (75 W glowing nichrome source), isomerization to the E-form was accelerated over 250 times that of the thermal isomerization and,

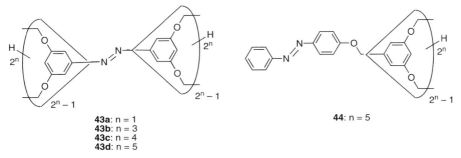

43a: n = 1
43b: n = 3
43c: n = 4
43d: n = 5

44: n = 5

Scheme 16 Dendrimer system reported by Jiang and Aida to exhibit low-energy photon harvesting [80–82]

remarkably, over 20 times that of the rate found on irradiation with visible light (440 nm). Yet, the rates of $Z \to E$ isomerization of **43a** and **43b** were unaffected by the infrared irradiation and, quite interestingly, UV radiation as well. Apparently, spatial isolation of the azobenzene is crucial for this effect, as indicated by control experiments involving compound **44**.

Infrared irradiation of Z-**43d** was carried out using three specific wavelengths: (a) a stretching vibration for aromatic rings ($1597 \, cm^{-1}$), (b) a stretching vibration for $CH_2 - O$ ($1155 \, cm^{-1}$), and (c) a transparent region ($2500 \, cm^{-1}$). Only the $1597 \, cm^{-1}$ radiation accelerates the $Z \to E$ isomerization reaction. In addition, irradiation of Z-forms of the entire series of dendrimers (**43a–d**) with 280 nm light (λ_{max} of the benzyl aryl ether dendrimer framework) accelerated the $Z \to E$ isomerization reaction, but only for **43c** and **43d** and not **43a** or **43b**. (This result is curious in that the azobenzene itself has a finite absorbance at 280 nm and irradiation at this wavelength markedly accelerates $Z \to E$ isomerization). These two results taken in concert strongly suggest a matrix to core intramolecular energy transfer is partly responsible for the acceleration effect. Hence, the authors postulate that the dendrimer frameworks in **43c** and **43d** insulate the interior units from collisional energy scattering as well as serve as light harvesting antenna. Photon harvesting is necessary to account for how $1597 \, cm^{-1}$ light (0.2 eV) could accelerate a process that has an activation free energy ($\Delta G^{\ddagger}$) of $19.4 \, kcal \, mol^{-1}$ (0.84 eV) at 21 °C. Indeed, photon flux experiments indicated that 4.9 photons at $1597 \, cm^{-1}$ (0.98 eV total) are involved in this photochemical process.

Porphyrins were chosen as an alternative chromophore to elucidate whether this selective energy transfer was a feature of just the azobenzene chromophore or the morphology of the entire dendrimer [66]. A variety of symmetrical and unsymmetrical dendrimers were synthesized in which the four *meso*-positions on the porphyrin molecule were substituted with methoxy-terminated poly(aryl ether) dendrons (3rd or 4th generation) and tolyl groups (Scheme 17). When **45e** was excited at 280 nm (dendron absorption), emission was observed primarily from the porphyrin core (656, 718 nm) with only weak emission seen from the dendrons (310 nm), giving an energy transfer quantum yield, Φ_{ENT}, of 80.3%. However, excitation at the same wavelength (280 nm) of the unsymmetrical dendrimers, (**45b–d**), resulted in emission primarily from the dendrons and only weakly from the porphyrin cores, yielding Φ_{ENT} values of 10.1, 19.7, 10.1, and 31.6%, respectively. This increase in Φ_{ENT} with increasing number of dendrons suggests a cooperativity of dendrimer subunits involved in the energy transfer process, and further indicates the dependence of the process on morphology of the dendrimer. Fluorescence anisotropy experiments show that once excitation of the dendrons occurs in **45e**, the energy migrates freely among the dendrimer subunits until it is transferred to the porphyrin core. However, in the unsymmetrical dendrimers, energy migration is less efficient, and therefore energy transfer is lower. Temperature-dependent fluorescence measurements sug-

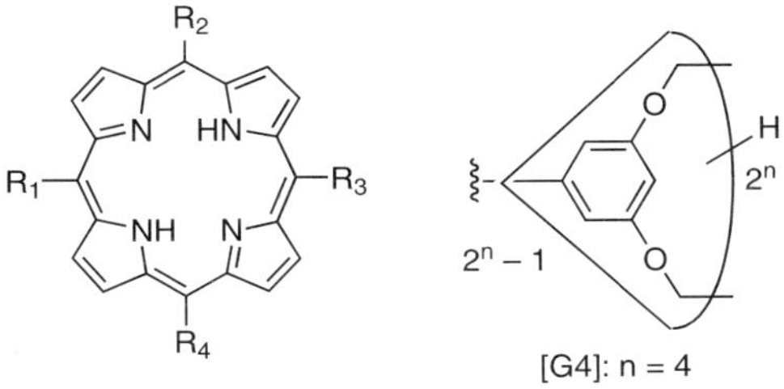

	R_1	R_2	R_3	R_4
45a	[G4]	tolyl	tolyl	tolyl
45b	[G4]	[G4]	tolyl	tolyl
45c	[G4]	tolyl	[G4]	tolyl
45d	[G4]	[G4]	[G4]	tolyl
45e	[G4]	[G4]	[G4]	[G4]

Scheme 17 Schematic of dendrimers containing a single porphyrin core by Jiang and Aida [66]

gest that lower energy transfer for the unsymmetrical dendrimers is related to conformational flexibility. Increasing temperature results in lower Φ_{ENT} values, yet the Φ_{ENT} of **45e** remains essentially the same up to 80 °C.

In a separate study by Shirai and coworkers, porphyrin dendrimers containing conjugated poly(phenylene) dendrons were synthesized (Scheme 18) and exhibit Φ_{ENT} of the highest generation of 98% [67]. This increase in efficiency over the poly(aryl ether) dendrons [66] was attributed to additional through bond (Dexter) energy transfer pathways from the conjugated dendrimer subunits to the porphyrin core. Fluorescence measurements on mixtures of [G2] boronic acid and the porphyrin core excited into the dendron absorption band (262 nm) exhibit emission only from the dendron. No energy transfer emission from the porphyrin appears. This again points to the important role that the dendrimer morphology plays in an energy-transfer process.

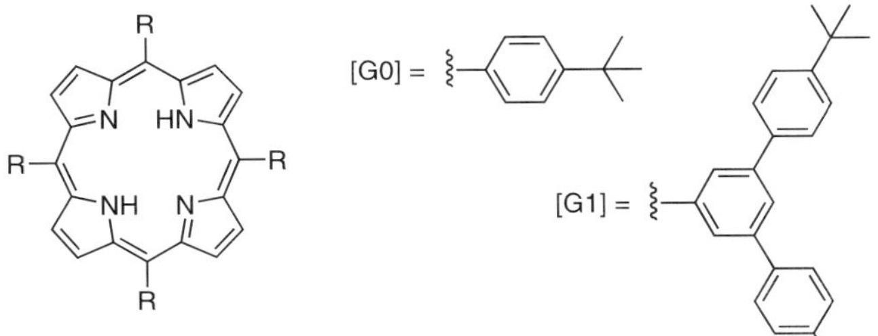

47a: R = [G1]
47b: R = [G2]

Scheme 18 Porphyrin dendrimers containing poly(phenylene) dendrons by Shirai and coworkers [67]

Fréchet and coworkers synthesized porphyrin cored dendrimers (Scheme 19) and their architecturally isomeric linear analogues which were then subjected to photophysical characterization [68]. Poly(aryl ether) linear oligomers were attached to a porphyrin core to give four-arm star and eight-arm star molecules. Corresponding eight-arm porphyrin cored dendrimers of generation 2 through 5 were exact architectural isomers, although their hydrodynamic volumes as measured by GPC vastly differed owing to the different conformations adopted by the linear and dendritic macromolecules. The dendrimer, with its more globular structure, gave a much lower hydrodynamic volume then its eight-arm isomer, whose molecular weight was closely predicted by linear polystyrene standards.

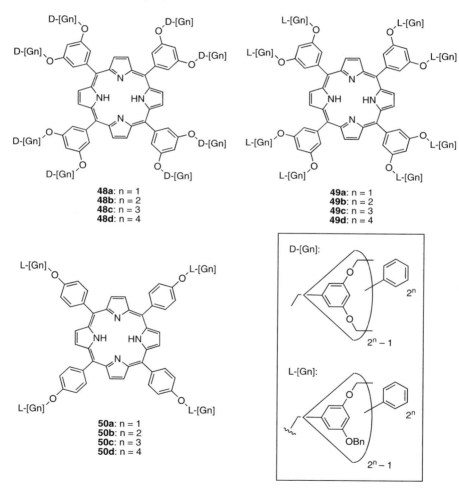

Scheme 19 Schematic of dendritic and four- and eight-arm porphyrins by Fréchet and coworkers [68]

To investigate the effect that linear versus dendritic poly(aryl ether) architectures might have on their antennae effect, emission studies were performed to elucidate Φ_{ENT} values. While the Φ_{ENT} only slightly decreased with increasing generation in the dendrimer series 89.7% for **48b**, 88.4% for **48c**, and 83.9% for **48d**, Φ_{ENT} for the 8-arm linear analogs falls off more dramatically (87.8% for **49b**, 74.1% for **49c**, and 57.0% for **49d**) corroborating the morphological dependence on Φ_{ENT} that was observed in previously described systems [66, 67]. While dendrimers provide a compact structure in which the distance between the dendron chromophores and the porphyrin core is within the Förster radius, the extended structure of the isomeric linear analogues can separate chromophores beyond a useful distance for energy transfer.

Goodson observed up-converted emission in a series of phenylazomethine dendrimers with a porphyrin core (Scheme 20) [69]. An investigation of the linear optical properties of these dendrimers revealed relatively low Φ_{ENT} values relative to other light-harvesting systems (36%, 30%, 18%, 17% for **51a–d**, respectively). These relatively low values were attributed to a combination of poor Coulombic coupling between the dendrons and the porphyrin core, reabsorption, dissipation of energy intramolecular vibrational energy redistribution (IVR), and singlet–singlet annihilation. Up-converted emission of the $Q_x(0,0)$ band (λ_{em} = 660 nm) was observed in all four generations upon excitation at 730 nm. The energy difference between the emission maximum and the excitation wavelength was $\sim 7\text{–}8$ times that of the thermal energy (207 cm^{-1}) of room temperature (kT), indicating a true frequency up-converted process. Since the power dependence of the fluorescence intensity did not show a quadratic relationship, a two-photon excitation process was ruled out. Hot-band absorption (HBA) was the more likely cause of the anti-Stokes luminescence (i.e. hot-band emission, HBE). The increase of the cross-section of HBA with an increase in generation number was attributed to enhanced intramolecular vibrational mode coupling in the dendrimer systems.

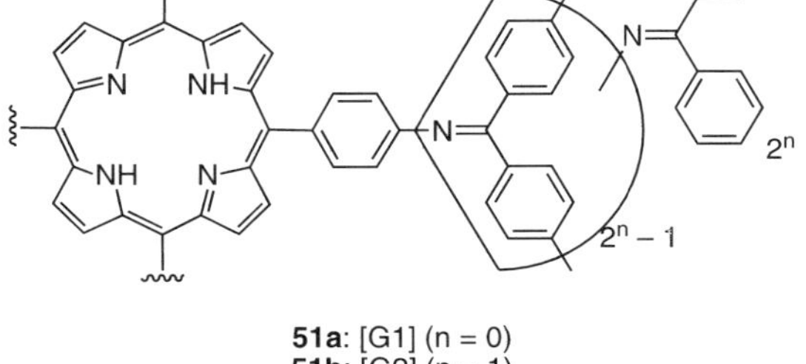

51a: [G1] (n = 0)
51b: [G2] (n = 1)
51c: [G3] (n = 2)
51d: [G4] (n = 3)

Scheme 20 Phenylazomethine dendrimers with porphyrin core by Goodson and coworkers [69]

5.2
Dendritic Porphyrin Arrays

Star-shaped multi-porphyrin arrays have been constructed to mimic energy funneling in photosynthesis. These dendrimers contain free-base porphyrin cores (P_{FB}) connected to four dendrons consisting of 1, 3, or 7 zinc-metallated porphyrins (P_{Zn}) embedded in an organic matrix connected by ether linkages (Scheme 21) [70, 71]. The periphery consists of methoxy-terminated poly(aryl ether) dendrons. For comparison, cone-shaped arrays consisting of P_{FB} cores monosubstituted with P_{Zn} dendrons were also synthesized.

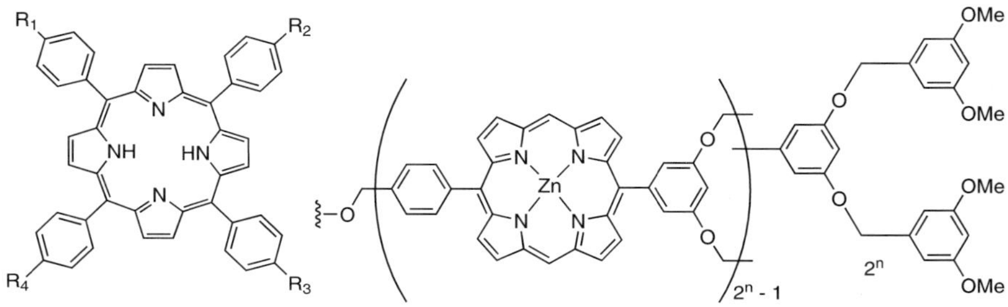

	R_1	R_2	R_3	R_4
52a	n = 3	n = 3	n = 3	n = 3
52b	n = 2	n = 2	n = 2	n = 2
52c	n = 1	n = 1	n = 1	n = 1
52d	n = 3	OMe	OMe	OMe
52e	n = 2	OMe	OMe	OMe
52f	n = 1	OMe	OMe	OMe

Scheme 21 Schematic of multiporphyrin array by Yamazaki and coworkers [70, 71]

Emission from the star-shaped arrays arises from excitation of the P_{Zn} followed by intramolecular energy transfer to the P_{FB} cores which act as energy traps. Measured Φ_{ENT} of the star-shaped arrays were 87, 80, and 71% for **52c**, **52b**, and **52a**, respectively. The drop-off of the cone-shaped arrays, however, was more dramatic with Φ_{ENT} of 86, 66, and 16% for **52f**, **52e**, **52d**, respectively. The large gap in Φ_{ENT} between **52e** and **52d** indicates that there is an upper limit to the distance between communicating chromophores when those chromophores are not conjugated, consistent with the Förster energy transfer model. Fluorescence lifetimes, τ_D, (monitored at 585 nm, an emission maxima of P_{Zn}) of the star-shaped arrays were always shorter then those of their cone-shaped counterparts. However, the gap between the star- and cone-shaped lifetimes became larger with increasing dendrimer generation, demonstrating the inability of the higher generation cone-shaped arrays to efficiently transfer energy to the P_{FB} core as corroborated by Φ_{ENT} values.

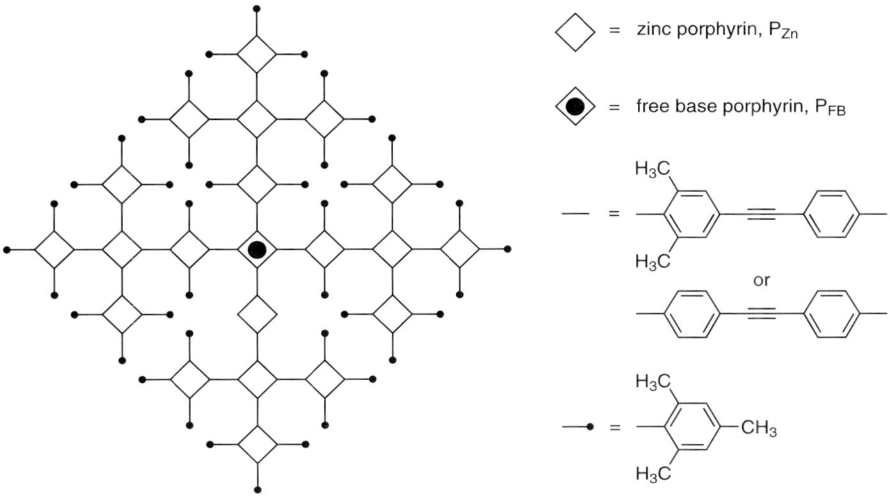

53

Scheme 22 Multiporphyrin array connected by diphenylethynyl linkages by Lindsey and coworkers [72]

Fluorescence anisotropies of the star-shaped arrays are lower then those of the cone-shaped molecules, which again indicates energy migration throughout the dendrimer framework followed by energy trapping by the P_{FB} core. Again, morphology of the dendrimer is critical for efficient light-harvesting porphyrin containing dendrimers as the dendritic subunits show cooperativity in both light-harvesting and energy funneling to the core.

Dendritic porphyrin arrays have been synthesized in which P_{Zn} molecules are connected via conjugated diphenylethynyl linkages and subsequently attached to a P_{FB} core (Scheme 22) [72]. These arrays show similar energy transfer characteristics to those mentioned previously [70, 71] (i.e. energy transfer occurs in a downhill fashion from the P_{Zn} to the P_{FB} core). However, the reported Φ_{ENT} values are much higher, presumably due to the increase in electronic communication by conjugation. Interestingly, bulk oxidation experiments show the removal of 20 electrons in **53**, which yields a stable π-cation radical with hole-storage possibilities.

5.3
Porphyrins as Peripheral Groups on Organic Dendrimers

Fifth generation poly(L-lysine) dendrimers have been modified at their periphery with 16 P_{Zn} in one hemisphere and 16 P_{FB} molecules in the other by using the orthogonal protection offered by Boc and Fmoc chemistry (Scheme 23) [73]. Energy transfer from the P_{Zn} to the P_{FB} was measured at

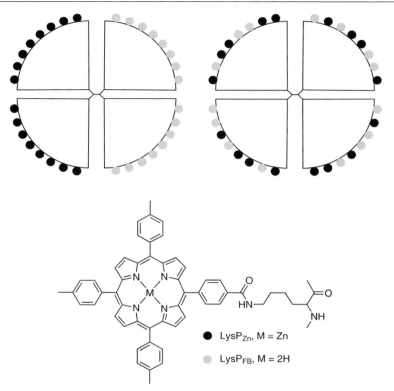

● LysP$_{Zn}$, M = Zn

◔ LysP$_{FB}$, M = 2H

Scheme 23   Porphyrins on the periphery of poly(L-lysine) dendrimers by Nishino and coworkers [73, 74]

43% which corresponds to roughly 7 of the 16 P$_{Zn}$ being in proximity to the P$_{FB}$ groups. When the porphyrins are arranged in a random fashion on the periphery (not necessarily 16 of each), energy transfer is increased to 85% [74].

Meijer and coworkers have synthesized a family of poly(propylene imine) dendrimers of first, third, and fifth generations that contain 4, 16, and 64 P$_{FB}$ moieties, at the periphery, respectively (Scheme 24) [75]. Fluorescence anisotropy values led to mechanistic models and equations to describe the energy transfer processes. The models take into account the different morphologies of the dendrimers: the disk-like nature of [G1] and the spherical natures of [G3] and [G5]. In the [G1] dendrimer, the Förster mechanism is sufficient to explain the energy transfer. Low anisotropy values in the [G3] dendrimer are explained by considering the dendrons as electronically separate from one another. Intradendron porphyrin energy transfer results in limited energy migration and therefore low anisotropy values. A segregated dendron explanation is not feasible for the crowded periphery of the [G5] dendrimer, as an intricate anisotropy decay indicates. The initial decay is fast

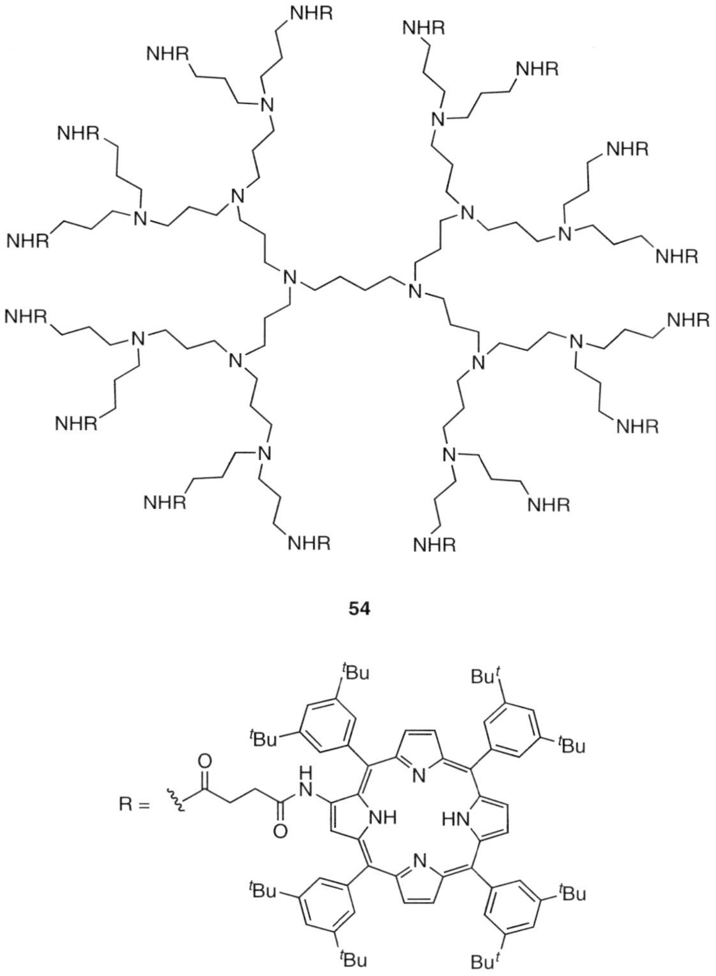

54

Scheme 24 Porphyrins at the periphery of PI dendrimers by Meijer and coworkers [75]

followed by a slow return and leveling off. Rapid energy transfer between the dense porphyrins at the surface of the sphere in addition to slower energy transfer between porphyrins further removed from the sphere can explain the observed anisotropy.

5.4
Porphyrin Core with Donor Periphery

Onitsuka and Takahashi reported the synthesis and characterization of zinc porphyrin core dendrimers with platinum acetylide based dendrons

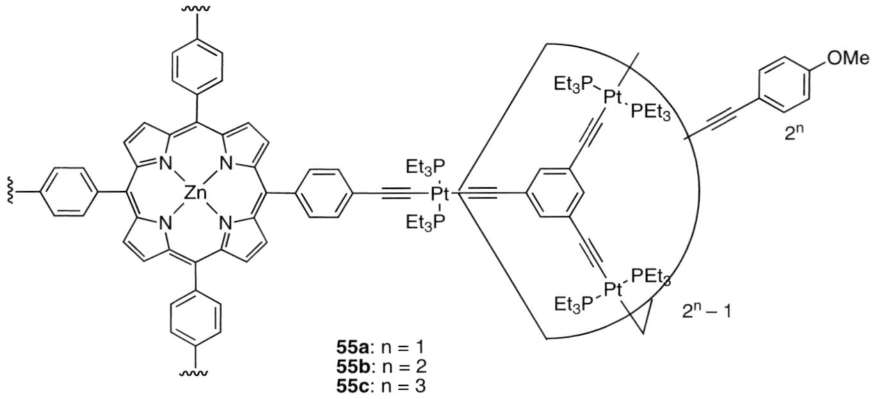

Scheme 25 Porphyrin core dendrimers with platinum acetylide dendrons by Onitsuka and coworkers [76]

(Scheme 25) [76]. Intramolecular energy transfer was evidenced by fluorescence of the porphyrin core observed (λ_{em} = 617 nm) upon excitation of either the MLCT band of the platinum acetylide dendrons (λ_{ex} = 344 nm) or the Soret band of the porphyrin core (λ_{ex} = 435 nm). Energy transfer from the dendrons to the core was not quantitative, as significant dendron emission (λ_{em} = 384 nm) was observed from all generations. Fluorescence from the porphyrin core at 617 nm appreciably decreased with increasing generation, and another fluorescence peak (648 nm) was observed.

A multichromophoric light-harvesting system with a free-base porphyrin core was constructed using a set of naphthopyranone dyes located on the interior and a set of coumarin chromophores located on the periphery of the dendrimer (Scheme 26) [77]. A similar pairing between coumarin donors and porphyrin acceptors had been studied previously by the same group [83, 84]. This assembly absorbs light over a broad range of the UV and visible spectrum and converts harvested photons into single emission from the central porphyrin chromophore via FRET. A lack of ground-state interactions among the dyes was verified by the absorption spectra that represented linear additions of the absorption spectra of the component chromophores. Near quantitave energy transfer was observed when compound **56** was excited at either of the two donor chomophore absorption bands (λ_{ex} = 335 and 358 nm).

Two similar systems with carbazole donor chromophores and porphyrin core acceptors were reported by the groups of Dehaen [78] and Lu [79]. With varying linker architecture and branching methods, Dehaen's system included dendrimers with 4, 8, and 12 carbazole chromophores per porphyrin core (Scheme 27). Lu's system also included materials with 4, 8, and 12 carbazole chromophores per porphyrin core, but with a different branching architecture (Scheme 27). The absorption spectra of the dendrimers in both of these studies are additive, with the visible transitions of the porphyrin cores,

56

Scheme 26 Multichromophoric dendrimer with porphyrin core by Fréchet and coworkers [77]

i.e. the intense Soret band at ca. 425 nm and the weaker Q-bands at longer
wavelengths, and of bands attributed to the carbazole subunits in the UV region. Emission in Dehaen's dendrimers (e.g. **57**), both at 77 K in a rigid glass
and at room temperature in fluid solution, was typical of porphyrin chromophores and independent of the excitation wavelength ($\lambda_{ex} = 330{-}630$ nm).
This, along with the strong overlap of the excitation and absorption spectra, indicates quantitative energy transfer from the carbazole chromophores
to the porphyrin core. Excitation of Lu's dendrimers (e.g. **58**) in dilute solution into either the Soret band ($\lambda_{ex} = 432$ nm) or the carbazole region
($\lambda_{ex} = 293$ nm) resulted in emission from the porphyrin core ($\lambda_{em} = 664$ and
730 nm), along with residual carbazole emission ($\lambda_{ex} = 394$ nm). Indeed, the
Φ_{ENT} values obtained by comparing the absorption spectrum and excitation

57

58

◀ **Scheme 27** Porphyrin core dendrimers with carbazole dendrons by Dehaen [78] (*top*) and Lu [79] (*bottom*)

spectra recorded at 293 nm were 69%, 65%, and 40% for the first, second, and third generation dendrimers, respectively.

6
Coumarin Dye Labeled Poly(aryl ether) Dendrimers

Fréchet and coworkers [85, 86] synthesized a series of dendrimers whose energy transfer mechanism is exclusively through-space. By designing dendrimers in which the donor periphery chromophores are effectively separated from the interior acceptors, the dendrimer architecture becomes simply a structural scaffold upon which chromophores can be placed. Chromophores were carefully chosen to satisfy the requirements of Förster energy transfer (i.e. emission of donor overlaps absorbance of acceptor), so that any photons absorbed by a molecule on the periphery undergo intramolecular singlet energy transfer to the core moiety and emission from that core ensues.

In these studies a pair of coumarin dyes were employed as the donor/acceptor pair. They were used for reasons that include: commercial availability at high purity, solubility in organic solvents, high fluorescence quantum yield (Φ_{fl}), sufficient spectral overlap (Fig. 6), and a large Stokes shift of the peripheral coumarin 2 dye ensuring that energy transfer, rather then self-quenching will be more probable following excitation.

To accommodate the nucleophilic nature of the coumarin 2 periphery dyes, "reverse" dendrons [33] were used so that the dendrimer could be convergently synthesized (Scheme 28). "Reverse" refers to using 3,5-benzyl groups versus the typical 3,5-phenols, thereby switching the reactivity at the 3 and 5 positions from nucleophilic to electrophilic making them reactive towards the nucleophilic coumarin 2. The interior dye was coumarin 343 (C343), whose absorbance overlapped sufficiently with the emission of coumarin 2.

Energy transfer in the first through third generation dendrimers (**59a–c**) is nearly quantitative as measured by comparing absorbance and excitation spectra and by studying fluorescence quenching of the donor by the acceptor. For the fourth generation (**59d**), the energy transfer efficiency decreases to 93% which is likely due to the increased interchromophoric distance. Energy transfer to the dendrimer scaffold is unlikely because its excited state lies at higher energy than both the coumarin donor and acceptor.

The significance of the dendrimers as effective light harvesters is most apparent in their "amplified emission." "Amplified emission" occurs when the emission intensity from the core is greater when the donor is excited rather then the core itself, and is a direct result of both the light harvesting abilities of the donors and the energy transfer efficiency to the acceptor. Compari-

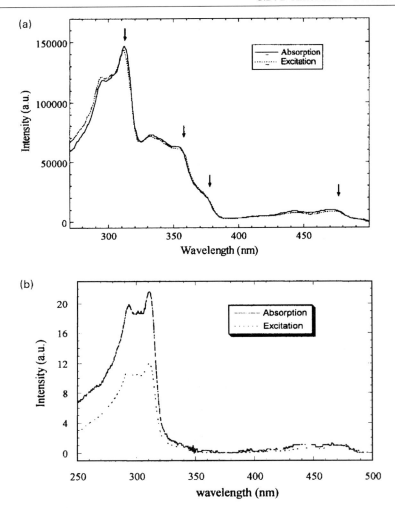

Fig. 6 Absorption and excitation spectra of an **a** "extended" nanostar and **b** "compact" nanostar [48]

son of fluorescence spectra indicate that there is a pronounced "amplified emission" for the higher generation dendrimers **59c** and **59d**, which contain 8 and 16 donor chromophores, respectively (Fig. 7). However, the increase in emission intensity does not scale with the absorbance increases for higher generations. The authors attribute this to an increase in non-radiative pathways for relaxation of excitons. Several generations of dendrimers containing coumarin-2 donor dyes on the periphery of "reverse" dendrons with penta- and heptathiophene acceptors at the core were prepared and also exhibited similar energy harvesting capabilities to the coumarin-2/C343 systems [87].

59a: n = 1
59b: n = 2
59c: n = 3
59d: n = 4

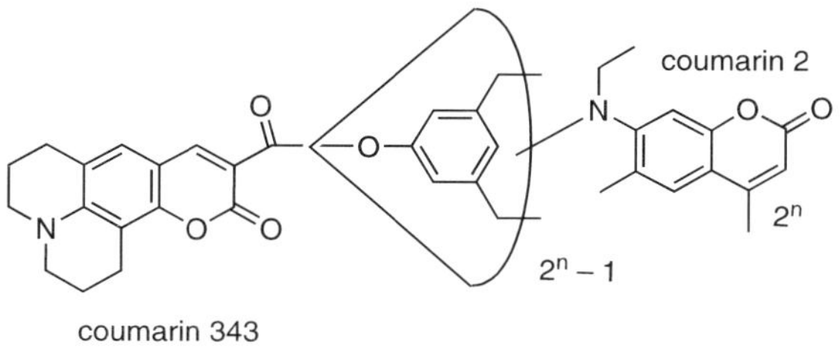

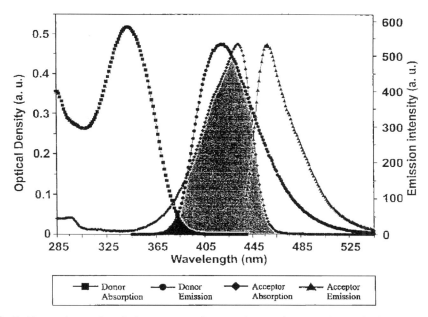

Scheme 28 "Reverse" dendrons with coumarin 2 hosts and coumarin 343 guest by Fréchet and coworkers [85, 86]

Fig. 7 Absorption and emission spectra of coumarin 2 and coumarin 343 [86]

Dendrimers containing chromophore donor/acceptor pairs carefully placed on an inert scaffold have found a use as the emitting material in light emitting diodes [88, 89]. If the donor/acceptor molecules also have hole and/or electron transporting capabilities, then a single layer device is possible. Fréchet and coworkers have synthesized dendrimers with hole-transporting triarylamines on the periphery and an energetically overlapping emissive acceptor on the interior (Scheme 29) [88, 89]. These dendrimers, when mixed with electron transporting oxadiazole and sandwiched between electrodes, emit

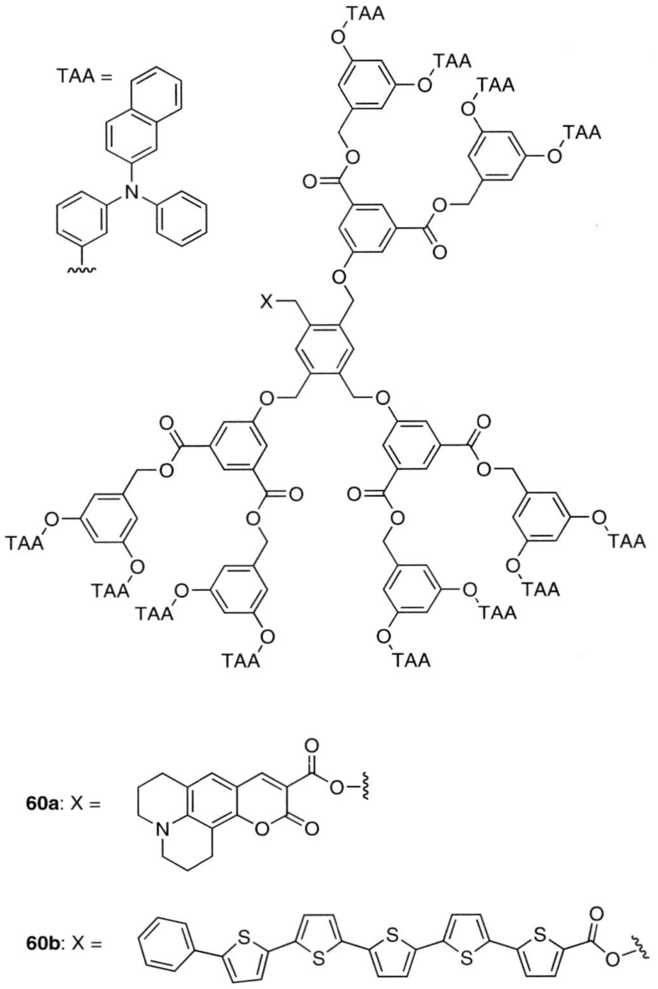

Scheme 29 Dye-labeled dendrimers with triarylamine peripheries for OLEDs by Fréchet and coworkers [88, 89]

light with external quantum efficiencies of 0.012% using C343 as the emitter, and 0.12% when pentathiophene (T5) is used as the emitter. Mixed dendrimer films (both T5 and C343 dendrimers) showed light emission primarily from the T5 (lower energy) with only a small emission from the C343, indicating that energy transfer also takes place between the emitting groups in neighboring dendrimers since the Förster radius (35–38 Å) is smaller then the interchromophoric distance. Energy transfer is not quantitative, however, since the emission of light from the C343 dendrimers increases linearly as the percentage of C343 dendrimer increases in the blend.

To overcome this large Förster radius and create mixed dendrimer multi-color OLEDs, larger dendrimers are needed so that energy transfer does not occur. However, poor solubility and increasing crystallinity of the dendrons prevented the synthesis of these target dendrimers [88, 89]. To overcome these problems, Fréchet and coworkers [90, 91] replaced every other triarylamine group on the periphery of the dendrons with a dialkyl substituted phenyl ring (Scheme 30). Dendrimers up to the 4th generation could be synthesized with C343 cores (e.g. **61**), and up to the 5th generation for T5 cores when only two dendrons are attached to either end of the T5 core in a "barbell" fashion (e.g. **62**). Site isolation of the individual cores, which implies that interchromophoric energy transfer is prevented, was probed by monitoring emission of C343 in films (thickness 1100–1300 Å) of mixtures of the dendrimers by photoluminescence (PL) and electroluminescence (EL). Results indicate that while site isolation for the C343 dendrimers is considerable at the fourth generation, it is only when the T5 dendrimers reach the fifth generation that their site isolation is significant. This is not surprising when considering the shape of the dendrimers. Presumably, the three-fold architecture of the C343 structures surrounds or encapsulates the core more effectively then the "dumbbell" architecture in the T5 dendrimer.

Because of the site-isolation afforded by the higher generation dendrimers, there is a large increase in C343 emission (normalized to T5 emission) when the molar ratio of C343 to T5 is increased from 1 : 1 to 5 : 1, indicating a diminishing or lack of energy transfer. Upon dilution of mixtures of [G4] C343 and [G5] T5 by embedding in a polystyrene matrix, emission of C343 again increases, due partly to the reduced energy transfer to the T5 dendrimers but mostly to reduced self-quenching. The experiment also indicates that even at high generations of both the C343 and T5 dendrimers, complete site-isolation is still not achieved, although there is a significant improvement over lower generations. OLEDs were fabricated and showed external quantum efficiencies of 0.2% for mixed dendrimer devices and 0.76% for [G5] T5 devices alone. The higher efficiency for the [G5] T5 is attributed to its ability to trap electrons more effectively then C343 dendrimers which leads to exciton formation and subsequently light emission. Note the higher efficiencies then previously reported [88, 89] (0.012% for C343 emitters and 0.12% for T5 emitters) with lower generation dendrimers.

Coumarin 2 has been paired as a donor with a diamino-substituted perylene near-IR emitter to produce a FRET-based UV to NIR frequency converter (Scheme 31) [92]. Excitation at the coumarin 2 λ_{max} ($\lambda_{ex} = 345\,nm$) resulted in FRET to the higher excited state (S_n) of the perylene core. This state undergoes rapid internal conversion to the first excited singlet state (S_1, Kasha's rule), from which emission is observed. Comparison of the integrated donor emission in the absence of the core and in the target dendrimer indicated a 99% energy transfer efficiency, accompanied by a 6.2-fold increase in the core emission relative to the emission in the absence of peripheral donors.

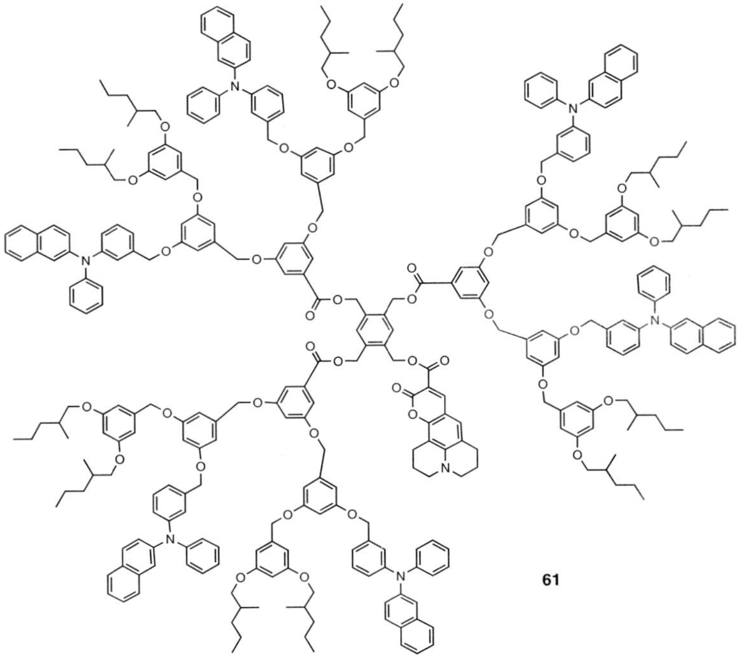

Scheme 30 Dye-labeled dendrimers with mixed peripheries by Fréchet and coworkers [90, 91]

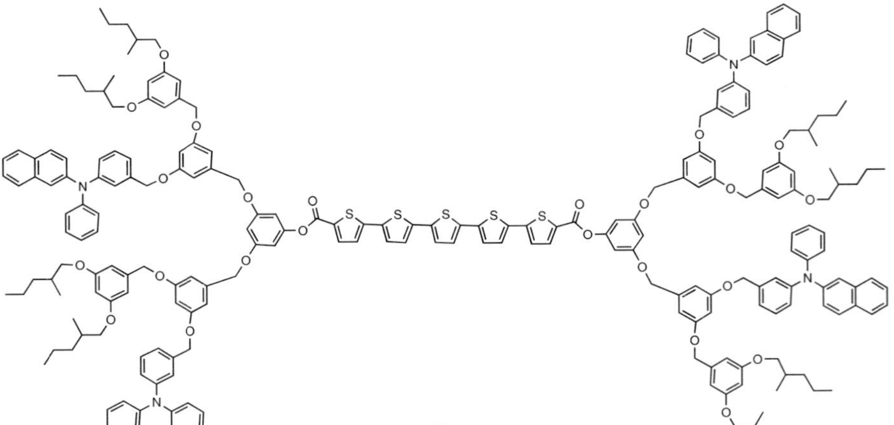

Fréchet and coworkers [93] have also synthesized multi-chromophore cascade dendrimers whose excited state energies decrease towards the core. The authors have been interested in elucidating whether energy transfer occurs in a step-wise fashion to the lowest energy acceptor (**Ac**), or whether there

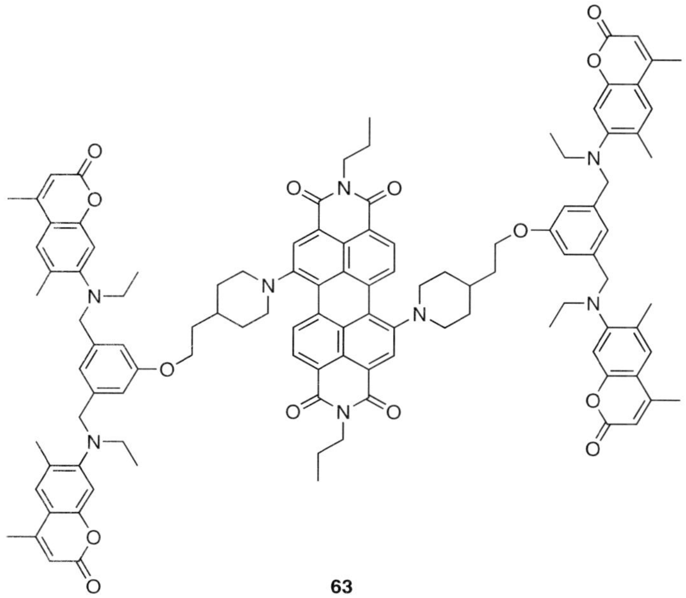

63

Scheme 31 FRET-based UV to NIR frequency converter dendrimer by Fréchet and coworkers [92]

is a component of energy transfer that takes place from the highest energy donor (**D1**) to **Ac** *bypassing* the middle chromophore (**D2**), even though **D1** and **Ac** were spaced farthest apart in the dendrimer array (Scheme 32). A series of model compounds were prepared combining only two chromophores so that energy transfer measurements could be made. FRET efficiencies for **D1** to **D2** (62; 99%), **D1** to **Ac** (63; 79%), and **D2** to **Ac** (64; 96%) were measured by comparing the emission of the donors with and without the presence of acceptors. Since the measured FRET in the dendrimer containing both donor chromophores, **D1** and **D2** (61), and the perylene acceptor, **Ac**, at the core was greater then 95% when exciting **D1** (342 nm), a cascade energy transfer from **D1** to **D2** to **Ac** seems likely since **D1** to **Ac** FRET efficiency was much lower (79%) then that of the multi-chromophore dendrimer (95%).

Mixed self-assembled monolayers (SAMs) that consist of coumarin 343 acceptors and dendrons substituted with coumarin 2 donors at the periphery have been constructed on silicon wafers and their energy transfer properties investigated by front-face fluorescence spectroscopy [94] Energy transfer comparisons between mixed SAMs containing coumarin 2 donor dyes either in a dendritic array (65) or as a single molecule terminating a long chain mixed with acceptor C343 (66) shed light on the importance of the dendritic structure (Scheme 33). In the case of the dendritic array 65, energy transfer was efficient as noted from the lack of donor emission. Amplified emission

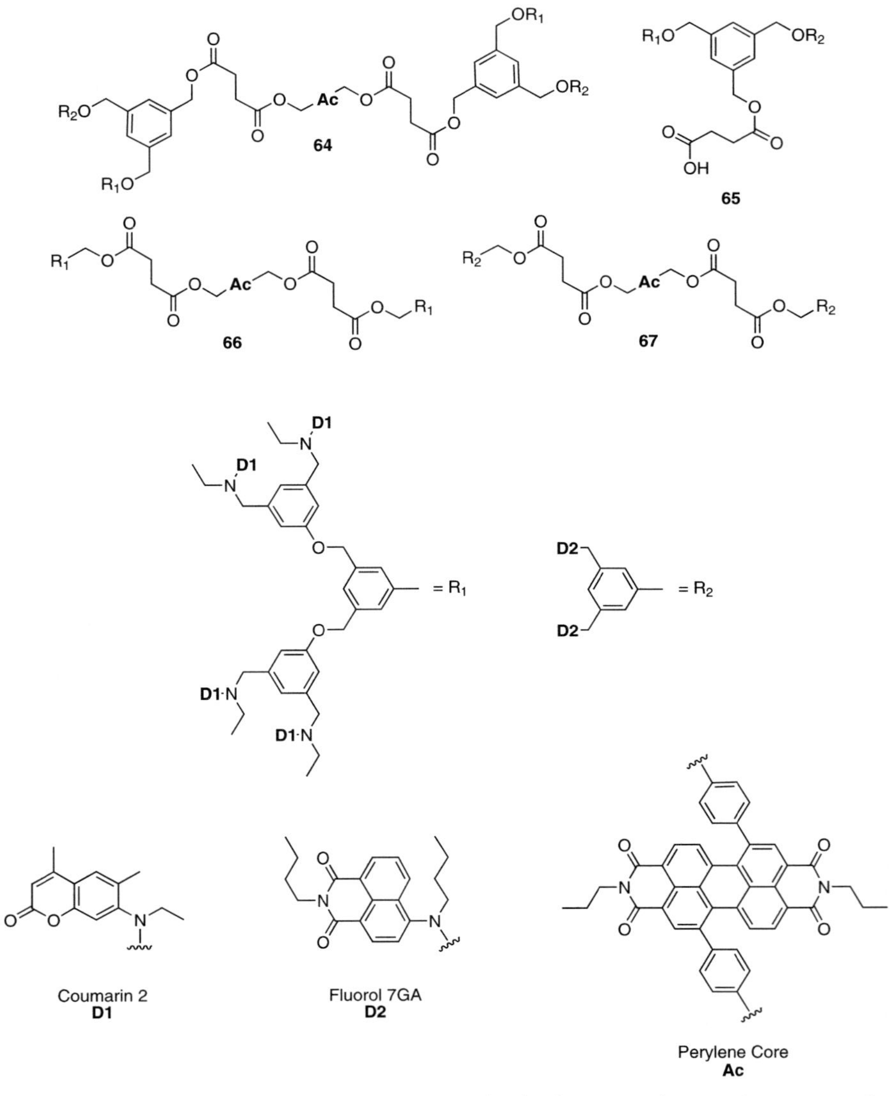

Scheme 32 Schematic of dendrimers with multiple donors and a single acceptor by Fréchet and coworkers [93]

was also apparent when the donors were excited rather then the acceptors. However, when the linear single molecules were mixed with the acceptors in a 4 : 1 ratio (to keep the molar ratio of donors to acceptors the same as in the dendritic case), there was significant donor emission. A likely explanation for the donor emission is that some chromophore pairs were too far apart (exceeding the Förster radius of 42 Å) due to phase separation of the donors

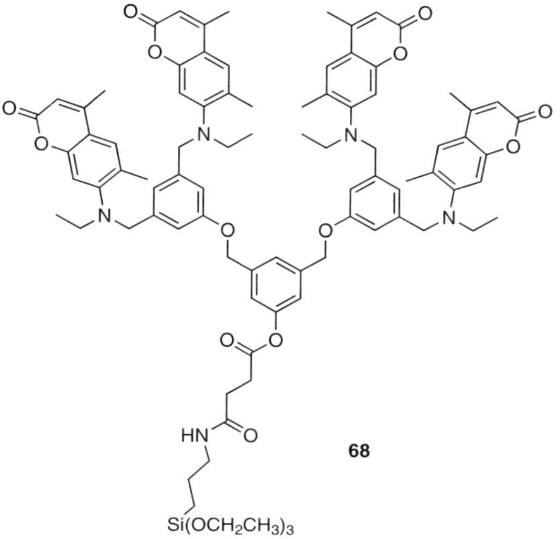

Scheme 33 Schematic of **68** and **69** for SAMs by Fréchet and coworkers [94]

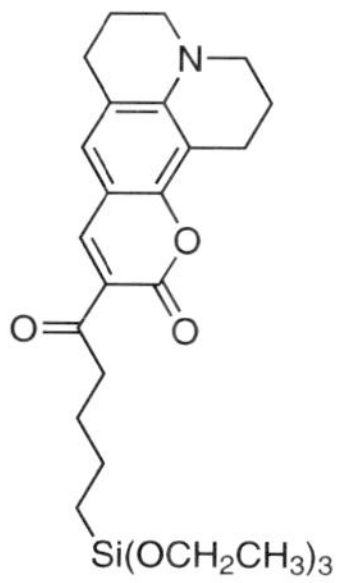

into domains. A dependence upon donor/acceptor molar ratio was observed in mixed SAMs of dendrons containing only two coumarin-2 dyes and linear acceptors. It was found that if the molar ratio exceeded 4 : 1, emission resulted from both acceptor and donor, but at 4 : 1, emission occurred almost exclusively from the acceptors. Lower molar ratios (1 : 2) also showed complete quenching of donor emission. However, no amplified emission was observed.

7
Rylene Cores

Perylenediimide and related moieties (i.e. rylene dyes) have been used as both donor and acceptor chromopores in energy transfer studies in dendrimers. These chromophores are generally photostable, have high extinction coefficients at convenient absorption wavelengths, and high fluorescence quantum yields. These chromophores have been incorporated most effectively into the shape-persistent polyphenylene class of dendrimers [95–102], although work has also been reported rylenes at the core of more flexible dendritic scaffolds [103–105].

7.1
Polyphenylene Dendrimers

In the past several decades, many different dendrimer architectures have been introduced for a variety of purposes. One of the simplest designs are the rigid polyphenylene dendrimers which consist only of branched phenyl groups. Since there are no bridging atoms between the rings, the phenyl groups of the dendrimer are twisted out of plane so that the dendrons themselves have limited conjugation. The bulky nature of the dendrons lends them as useful insulators for chromophores with aggregation, crystallinity, and solubility issues [95]. Also, as with other rigid dendrimer architectures, polyphenylenes can be used as inert scaffolds onto which host/guest chromophores can be placed for Förster energy transfer without the complication of backfolding of the dendritic branches [96–102]. These types of polyphenylene dendrimers have been studied extensively by Müllen and coworkers by a variety of techniques including fluorescence anisotropy and single-molecule spectroscopy.

The polyphenylene dendrimers were found to effectively isolate large chromophores such as perylene derivatives, when two polyphenylene dendrons (first through third generation) are attached to a perylenediimide core (PDI) (Scheme 34) [95]. Although the alkoxy substituents in the "bay" area of the perylene twist the core out of planarity, the chromophore is only slightly blue-shifted when incorporated into the dendrimers. Therefore, energy transfer from the polyphenylene dendrons is still accomplished with high efficiency.

When a phenylene first generation dendrimer with a tetrahedral core is substituted on the periphery with three peryleneimides (PI) and one terryleneimide (TI) (Scheme 35) to act as the lower energy acceptor, energy hopping between the PI molecules eventually results in energy transfer to the TI trap [96, 97]. Since the dendrimer only contains the four equally spaced chromophores, there are no intramolecular excimers formed, as evidenced by their localized fluorescence. Two different fluorescence anisotropy decay times measured for this system indicate that there are two different energy pathways available for trapping by TI. The authors postulate that there are two different environments for the TI groups with respect to the PI donors. One of the environments places the chromophores close to one another which provides the initial close-proximity pathway. The two pathways, however, do not affect the overall energy transfer efficiency which is notably high at 96%.

By changing the position of the chromophores so that the terrylenediimide (TDI) is at the core and the PI are at the periphery of polyphenylene dendrons, Müllen and coworkers have created dendrimers that are incapable of intramolecular aggregation of chromophores due to their rigidity (72 and 73) [98–101]. Additional high energy donors such as naphthaleneimides (NI) have been placed in a 2 : 1 ratio to PI on the periphery of the dendrons to create cascading dendrimers (74) which can undergo stepwise energy transfer from the NI to the PI and ultimately, to the TDI moiety at the core [98, 102]. Excitation of the PIs in the two-chromophore dendrimers (72 and 73) results in roughly 93% energy transfer to the TDI core in both 72 and 73, expected because of the large Förster interaction radium ($R_0 = 6$ nm) and good spectral overlab of the PI emission and TDI absorption. The multichromophoric triad system 74 absorbs over the whole range of the visible spectrum but

70

Scheme 34 Polyphenylene dendrimer with a perylenediimide core (PDI) by Müllen and coworkers [95]

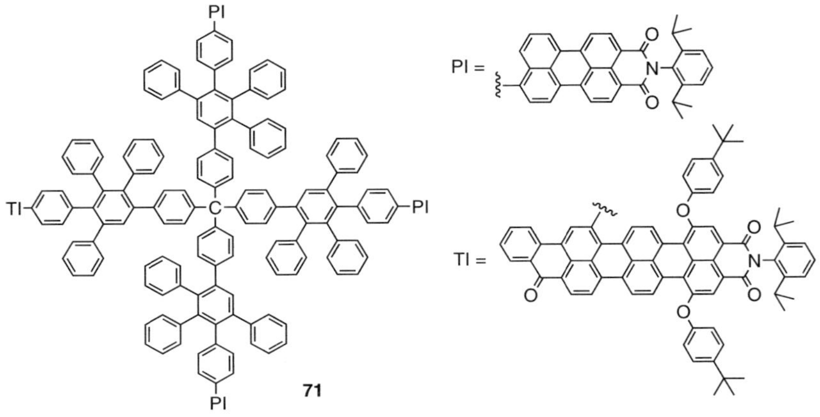

Scheme 35 Phenylene dendrimer with different rylene subunits by Müllen and coworkers [96, 97]

with well-separated absorption envelopes. Interestingly, energy transfer in the triad system **74** upon excitation of the NI chromophores (370 nm) occurs predominantly to the TDI (700 nm emission), despite an apparent lack of spectral overlap between these two chromophores [98]. However, further studies revealed a spectral overlap between NI emission and the S_0–S_2 absorption bands of the TDI chromophore, indicating the possibility of both direct transfer from NI to TDI as well as cascade transfer via the intermediate PI chromophores [102]. The diads **72** and **73** have also been studied at the single molecule level [99–101].

7.2
Flexible Dendrimers

Tian and co-workers have prepared and investigated the properties of flexible dendrons and dendrimers that have NI or PDI cores with carbazole (CZ) or oxadiazole (OXZ) peripheral units (Scheme 38) [103, 104]. Excitation of the peripheral chromophores result in emission of the NI of PDI cores. Interestingly, in the case of the NI core dendrons, excitation of the peripheral oxadiazole residues of **75c** results in a 3.9 times enhancement of core luminescence, while excitation of the peripheral carbazole residues of **76c** results in only a 20% emission intensity, suggesting a second pathway for donor quenching in the CZ systems. This pathway is probably photo-induced electron transfer (PET) from the CZ units to the NI core [104]. The OXZ units, however, have a relatively higher electron affinity and no PET can take place.

This PET pathway was studied directly in the analogous dendrimer systems with PDI cores (e.g. **77** and **78**) [103]. Direct excitation of the PDI cores (540 nm) in dendrimers bearing different numbers of OXZ units (e.g. 0, 2, 4)

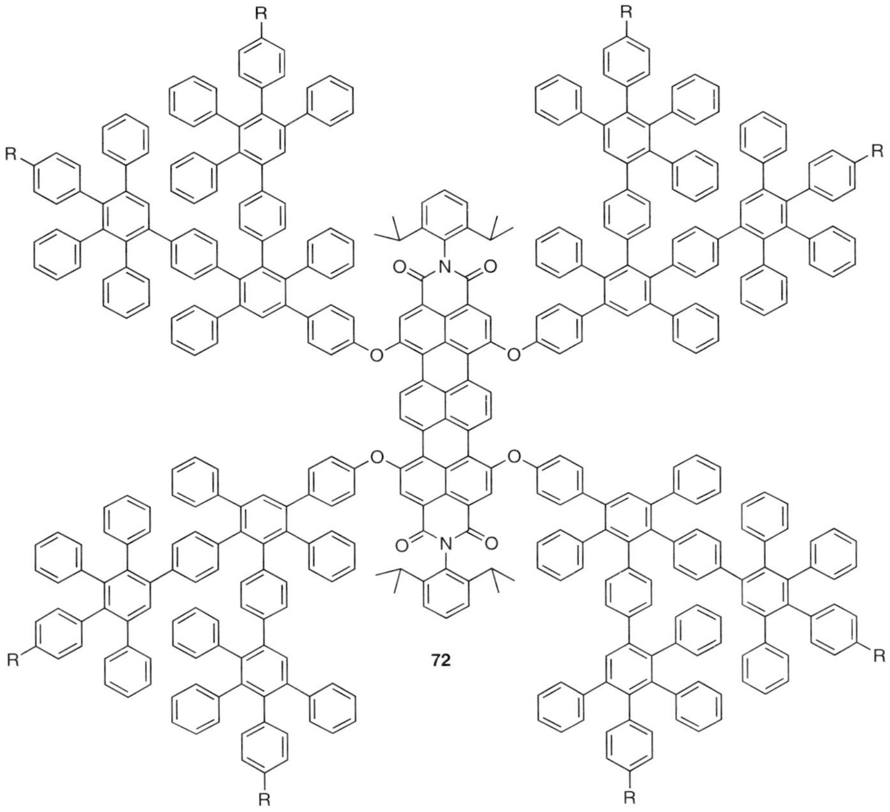

Scheme 36 Schematic of rigid polyphenylene dendrimers 72 and 73 [98]

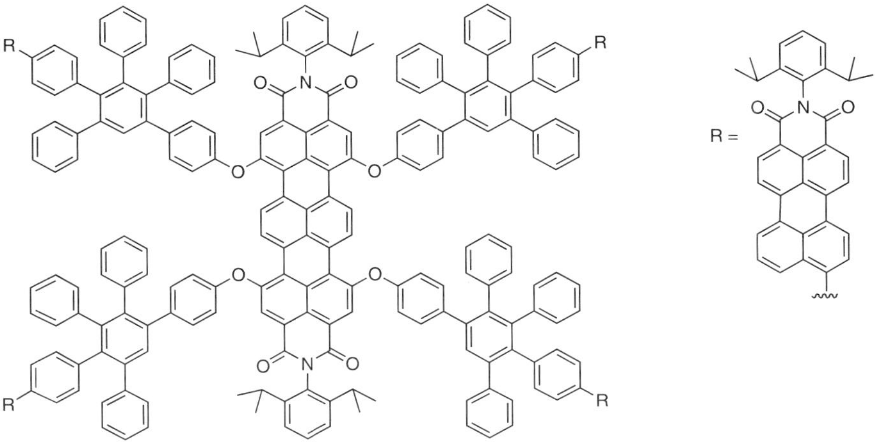

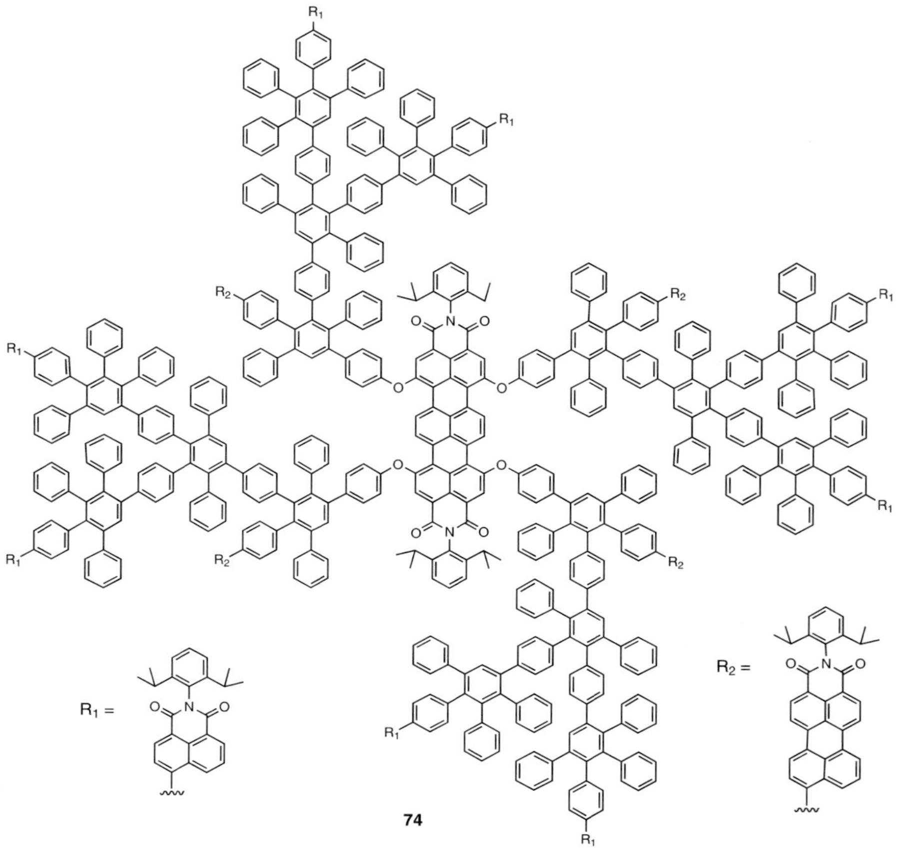

Scheme 37 Schematic of dendrimer 74 [98]

resulted in intensities of core emission that were the same within experimental error. However, the intensities for core emission of CZ-containing dendrimers under direct excitation of the core decreased sharply with increasing number of CZ units. The relative core emission intensities of dendrimers bearing 0, 2, 4, and 8 CZ units were 1.0, 0.48, 0.14, and 0.07, respectively. This is consistent with PET quenching between the core PDI and the peripheral CZ units. Hence, the predominant pathway following OXZ peripheral unit excitation is energy transfer to the PDI core and core emission, while for the CZ peripheral unit dendrimers, energy transfer to PDI is followed by core emission that competes with electron transfer from CZ to PDI and back electron transfer to the ground state. Based on fluorescence lifetime data, the authors calculate the PET efficiency to be in the range of ∼75%.

Takahashi et al. [105] reported a family of dendrimers with perylene tetracarboxylate cores and layers of 4, 8, 12, 16, or 24 anthracene units at-

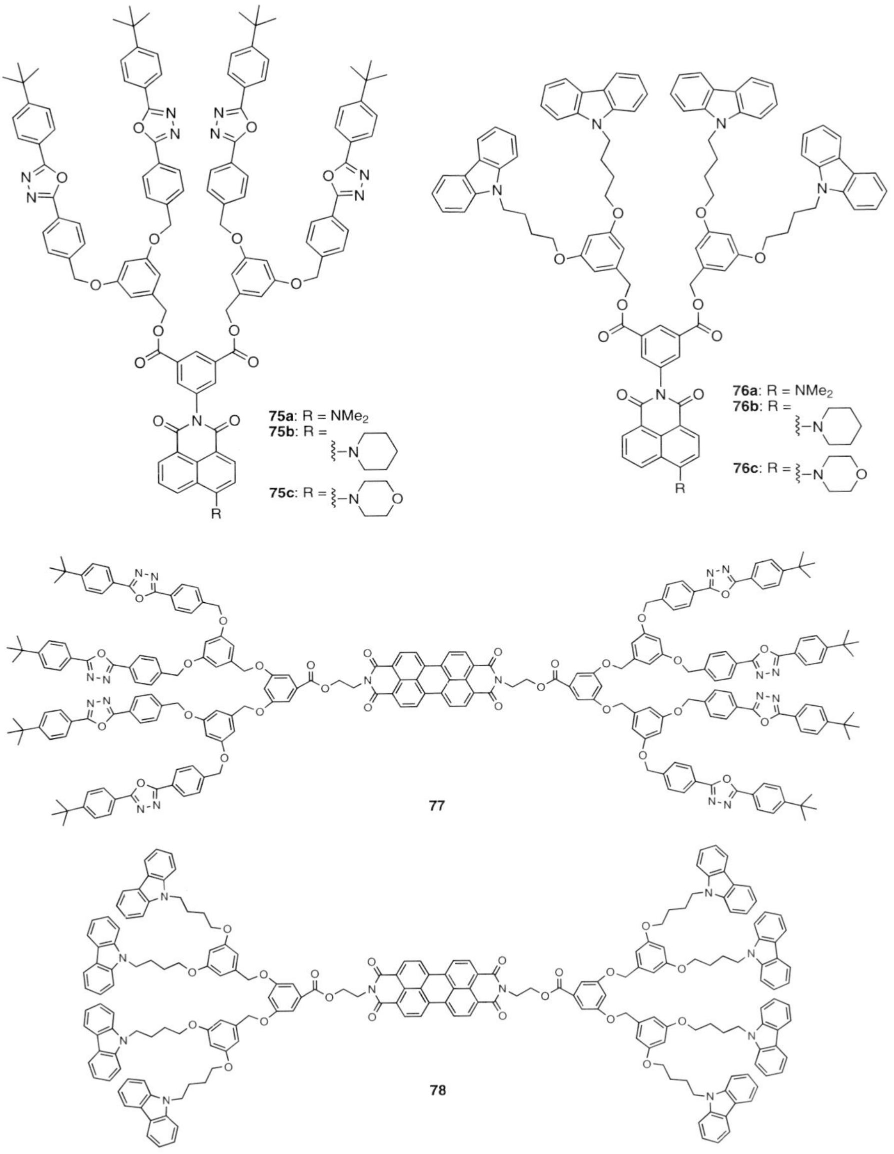

75a: R = NMe₂
75b: R =
75c: R =

76a: R = NMe₂
76b: R =
76c: R =

77

78

Scheme 38 Examples of NI and PDI core dendrons and dendrimers with CZ and OXZ donors by Tian and coworkers [103, 104]

tached through flexible benzyl based subunits. Evidence was obtained for decreased fluorescence efficiency due to intramolecular energy transfer in densely packed donor moieties, and quantitative energy harvesting by the Förster mechanism.

8
Two-photon Light Harvesting and Energy Transfer

Dendrons and dendrimers containing two-photon absorbing (TPA) chromophores as the donors and a Nile Red dye (NR) as the core acceptor have been synthesized for potential uses in two-photon imaging and optical limiting (Scheme 39) [106, 107]. These dendrimers were based on dendrons, reported by the same group, containing TPA chromophores at their chain ends that exhibited a linear correlation between the number of peripheral chromophores and the TPA cross section of the molecules [108] Single-photon absorption and energy transfer from the TPA to the NR in dendrimers consisting of one TPA and one NR resulted in a 3.4-fold increase in NR emission when exciting the TPA chromophores (405 nm) rather then the NR (540 nm) [106]. This can be attributed to the antennae effect of the TPA moieties which effectively increases the extinction coefficient of the molecule at the TPA single-photon absorbing wavelength (405 nm). Energy transfer for single-photon absorption was found to be greater then 99%. Laser-induced two-photon absorption also resulted in energy transfer from the TPA groups

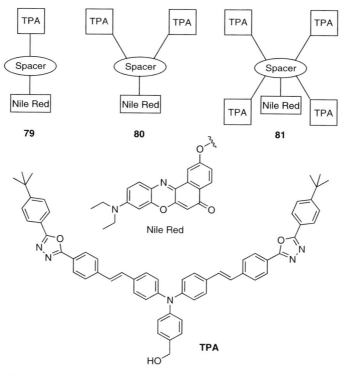

Scheme 39 Two-photon dendrimers with a Nile Red dye acceptor by Fréchet and Prasad [106, 107]

to the NR core [107]. Three dendritic compounds consisting of 1, 2, and 4 TPA hosts (**79, 80,** and **81,** respectively) surrounding a NR guest showed 8, 20, and 34-fold increases in NR emission, respectively, also attributed to the antennae effect. Compounds such as these which can absorb and transfer both single- and two-photon energy to a central core are important as energy harvesters because the window of wavelengths that can be used to excite the molecules is increased substantially over systems capable of only single-photon absorption.

An approach to singlet-oxygen generation sensitized by porphyrins at the core of dendrimers with containing donor TPA chromophores has recently been reported (Scheme 40) [109, 110]. Compound **82a** was prepared from eight AF-343 TPA donor chromophores and a multi-valent acceptor porphyrin and exhibited a steady-state absorption spectrum consistent with

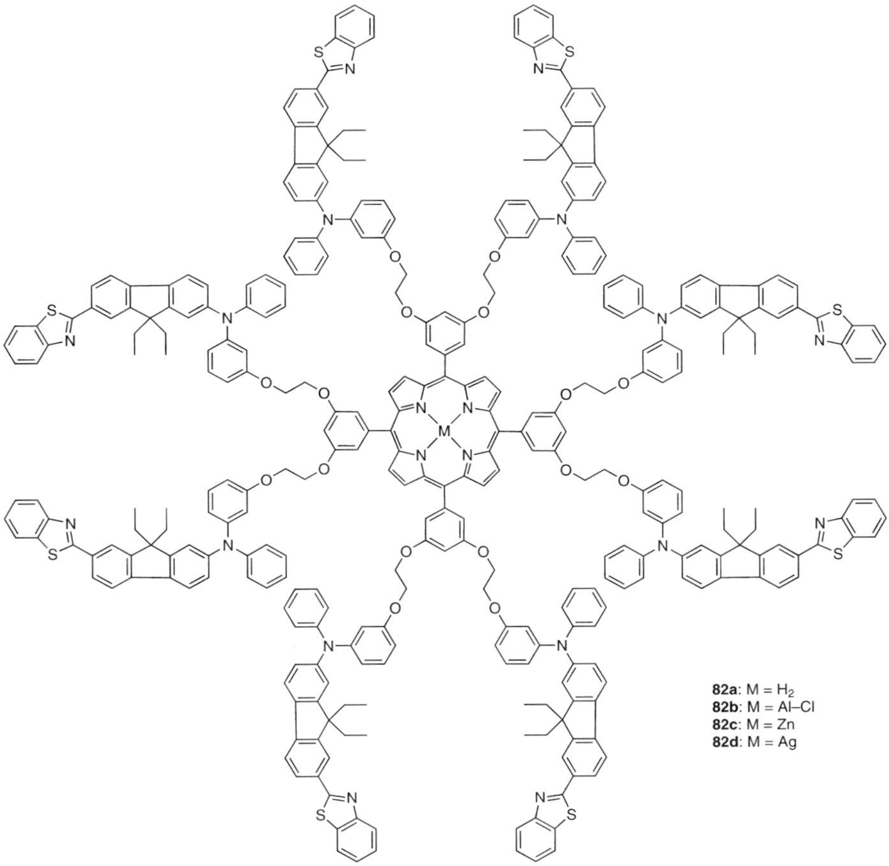

82a: M = H₂
82b: M = Al–Cl
82c: M = Zn
82d: M = Ag

Scheme 40 Porphyrin core dendrimers with TPA donors for singlet oxygen sensitization by Fréchet and Prasad [109, 110]

a linear addition of the absorption spectra of model compounds of the individual chromophores [110]. Single-photon excitation conditions at both the absorption maxima of the AF-343 chromophore (λ_{ex} = 385 nm) and the Soret band of the porphyrin core (λ_{ex} = 424 nm) resulted in porphyrin emission, with the energy transfer efficiency in the former case calculated at 97%. The emission spectrum of **82a** measured when a 780 nm fs mode locked Ti:Sapphire laser was the excitation source showed porphyrin emission 17 times more intense than emission from a monomeric model porphyrin under the same conditions. The ability of **82a** to generate singlet-oxygen following

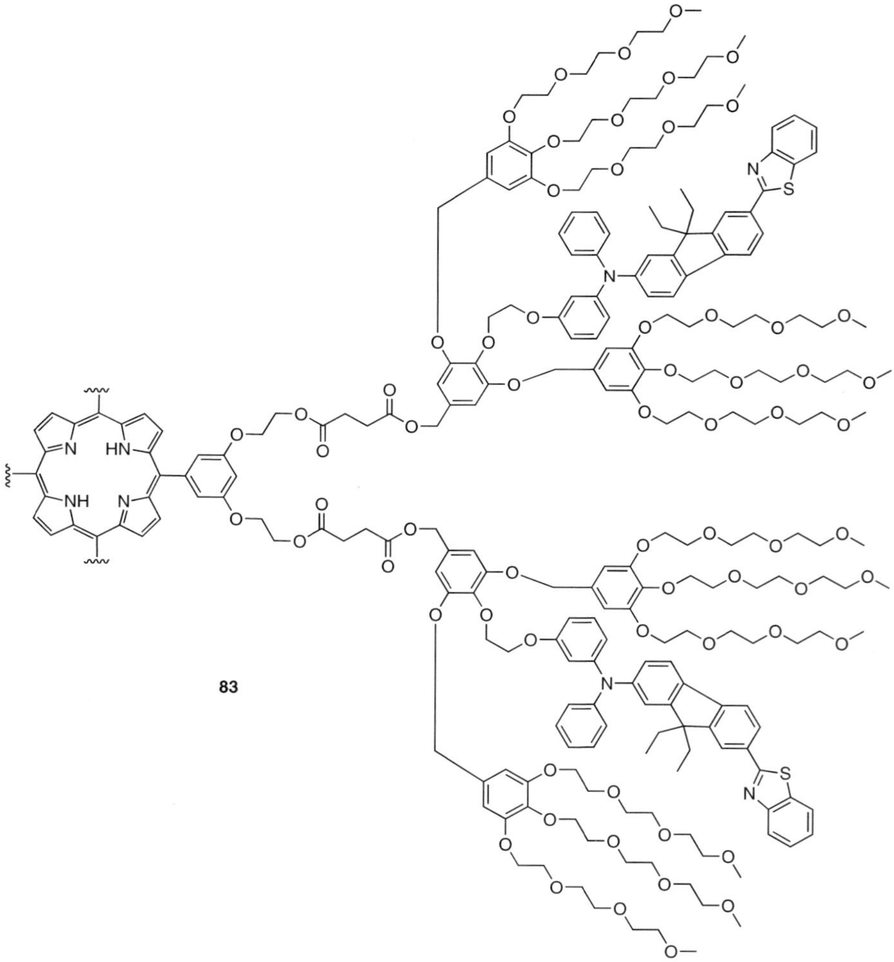

Scheme 41 Water soluble dendrimers with TPA donors for singlet oxygen sensitization by Fréchet and Prasad [111]

TPA was evaluated by monitoring the emission of singlet-oxygen at 1270 nm. Singlet-oxygen generation via TPA begins with absorption of two-photons to populate a higher that attained by one-photon excitation. Rapid internal conversion results in the first excited singlet state, which can then undergo FRET to the porphyrin, intersystem crossing to the triplet state, and then collisional deactivation by $^3O_2$ to produce $^1O_2^*$ and ground state porphyrin. Excitation at both single photon (532 nm) and TP (780 nm) wavelengths resulted in 1270 nm luminescence

Compound **82a** was metallated to produce compounds **82b**, **82c**, and **82d** to attempt to tune the singlet-oxygen generating efficiency of this system by increasing the efficiency of intersystem crossing to the triplet state [109] Time-resolved fluorescence, transient absorption measurements, and TP excitation experiments indicated efficient energy harvesting upon TPA and FRET to the central porphyrin chromophore for the Al–Cl and Zn species. The Ag complex was non-fluorescent, and whether the nature of the AF-343 quenching was via FRET or electron transfer was unclear. Metallation of the porphyrins strongly influenced the quantum yields of the triplet excited states for the Al–Cl and Zn species, and this increased triplet quantum yield resulted in enhanced singlet-oxygen production, as measured by singlet-oxygen emission intensities, although quantum yields for singlet-oxygen generation were not measured.

A water soluble version of this system with eight AF-343 TPA donor chromophores and a core acceptor porphyrin was also been produced (Scheme 41) [111]. Water solubility was achieved by inclusion of tri(ethylene glycol) (TEG) moieties in the periphery of the dendrimer. Although oxygen luminescence could not be observed upon TPE (lex = 780 nm) in D_2O solution, when the solvent was changed to benzene-d_6, weak emission was observed under TPE conditions.

9
Energy Transfer to Encapsulated Guests

While several researchers have investigated energy transfer from a host (often the dendron itself) to a guest at the core of the dendrimer, there are few examples of guests that are not covalently bound but encapsulated in some fashion into the voids of the dendrimer. In an elegant example by Meijer and coworkers, oligo (p-phenylene vinylene)s (OPPVs) are covalently attached to the periphery of a PPI dendrimer and transfer energy to an encapsulated anionic dye, Sulforhodamine B (Scheme 42) [112]. The dye is first loaded into the dendrimer by extraction from an aqueous solution into a dendrimer-containing organic phase. An acid-base reaction between the tertiary amines in the dendrimer and the carboxylic acids of the dye holds the dyes creates electrostatically bound guests. Seven dyes are loaded into the third generation

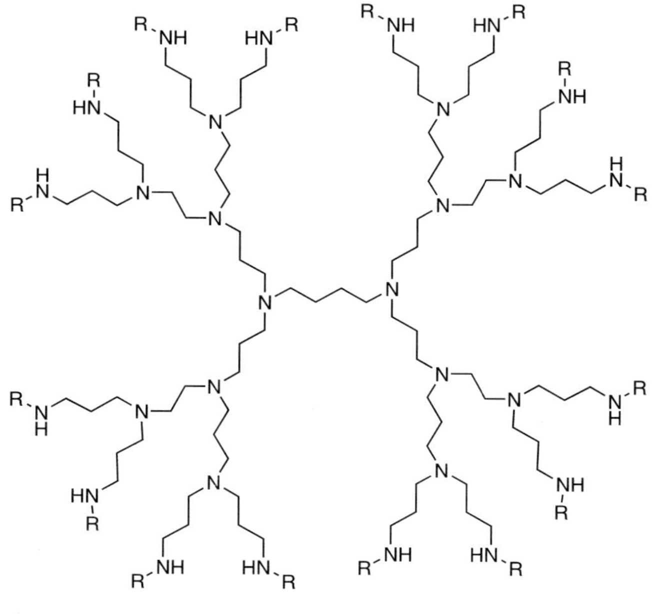

84

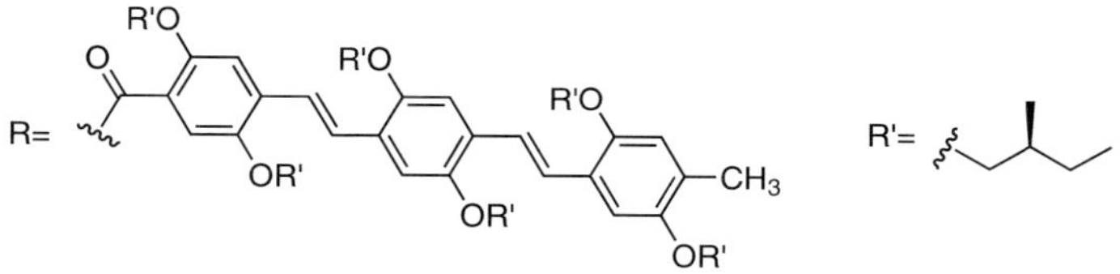

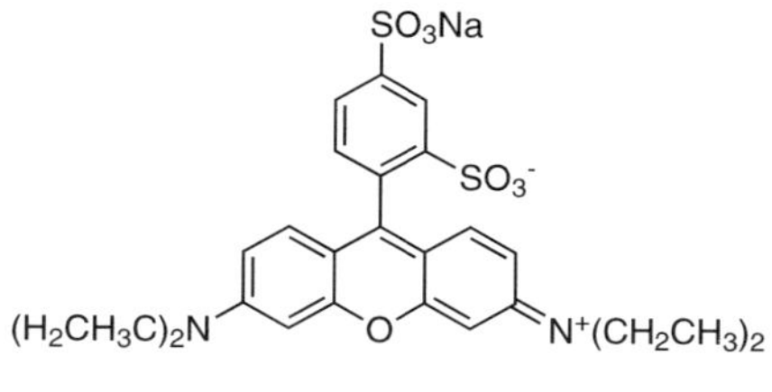

Sulforhodamine B

Scheme 42 Structures of third generation PPI-OPPV dendrimer **84** and Sulforhodamine B by Meijer and coworkers [112]

dendrimer **84**, while the fifth generation dendrimer is capable of extracting 26 dyes into its interior, occupying about half of the available tertiary amine sites located in each dendrimer. Solution photoluminescence measurements indicate that when the dye concentration in the dendrimer is at a maximum, energy transfer from the OPPV groups to the dye is approximately 40% for both third and fifth generation. Although the energy transfer percentage is lower than in most covalently incorporated guests, Meijer's system offers the unique opportunity of switching the wavelengths of emitted light if the dye can be removed from the dendrimer as easily as it is introduced.

Fluorescence titration curves show an initial steep increase in dye fluorescence when approximately one or two dyes are bound followed by only a gradual increase in fluorescence as each additional dye is bound to the interior of the dendrimers (Fig. 8). This change in slope could be due to either self-quenching of the dyes which will compete with fluorescence once the concentration of dye is high enough, or just simply an antennae effect of the OPPVs. Thin films of the dye-loaded dendrimers showed a much higher energy transfer efficiency (greater than 90%) which could be due to either better spectral overlap (the film fluorescence is slightly red-shifted) or better orientation of the host and guest molecules relative to each other within the film, a parameter in the Förster energy transfer equation. Additional dyes that were incorporated into the dendrimer all resulted in near-quantitative energy transfer in thin films.

Vögtle and Balzani have also explored light-harvesting and energy transfer to non-covalently linked guests [113–115]. They utilize PPI dendrimers modified with dansyl groups at the periphery (85) as the hosts for acidic dyes (i.e. eosin Y, fluorescein, rose bengal) as the encapsulated guest emitters (Scheme 43) [113, 114]. Dendrimers of second through fifth generation

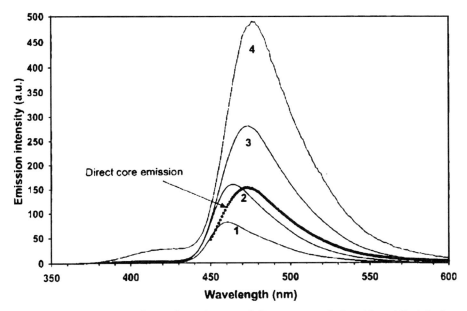

Fig. 8 Emission spectra of **59a–d** in toluene and direct core emission (*dotted line*) [86]

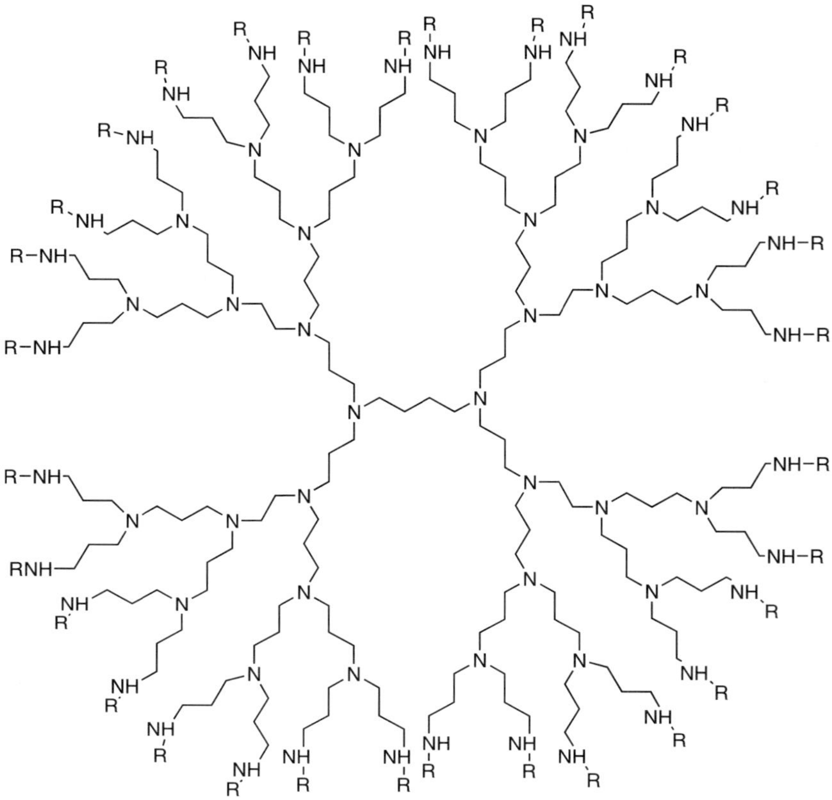

85

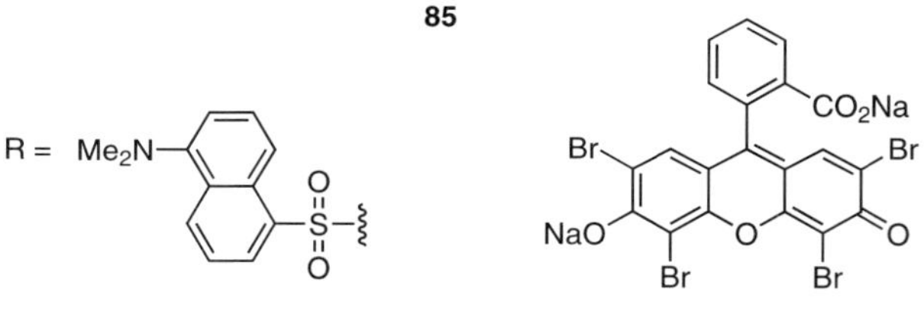

R = Me₂N— [dansyl group with S-O₂]

[Eosin Y structure with Br, CO₂Na, NaO, O groups]

Eosin Y

Scheme 43 PPI dendrimer with 32 dansyl groups at the periphery by Vögtle and Balzani [113, 114]

were capable of extracting eosin Y (a diacid) into their cavities. The amount of eosin molecules incorporated into the dendrimers was both pH and concentration dependent. Fluorescence measurements were performed mainly on the fourth generation dendrimer **85**, and show that when the 32 dansyl groups of the dendrimer are excited, the eosin guests quench their fluorescence. Emission from eosin, however, is also significantly quenched. This is thought to be due to non-radiative emission pathways made possible by the electrostatic bonds holding the eosin molecules in place. Encapsulation of one eosin molecule is enough to quench the fluorescence of all 32 dansyl groups.

A more recent example by Balzani and Vögtle provided an additional chromophore for cascade energy transfer [115]. Poly(aryl ether) dendrons with naphthyl groups at the periphery were attached to the sulfonamides of the second generation dansyl terminated dendrimers to create a "super" dendrimer **86** consisting of three different types of chromophores: 32 naphthyls, 24 alkoxybenzenes, and 8 dansyls (Scheme 44). When eosin is extracted into

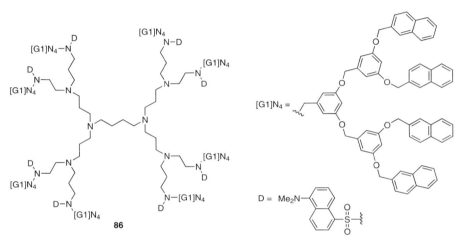

Scheme 44 Schematic of "super" dendrimer **86** by Vögtle and Balzani [115]

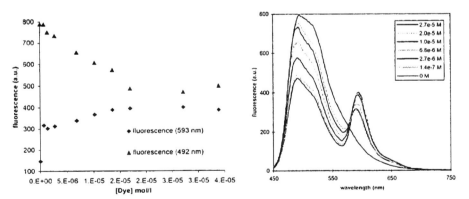

Fig. 9 a Fluorescence titration curve monitoring both the emission of the G3 dendrimer (492 nm) and the Sulforhodamine B dye (593 nm) in the CHCl$_3$ layer as the concentration of dye is increased in the water layer. **b** Fluorescence spectra of the CHCl$_3$ layer as concentration of the dye is increased in the water layer [112]

the interior of the dendrimer and the dansyl groups are excited, emission occurs from eosin with greater then 80% energy transfer efficiency. Any light harvested by the naphthyl or dimethoxybenzyl chromophores is thought to undergo energy transfer to the dansyl units and then on to the eosin guest(s). Therefore, a cascading energy transfer dendrimer was realized.

10
Conclusion

Dendrimers have been shown to act as efficient light harvesters. Since they can be easily synthesized and modified to include chromophores, many routes to light-harvesting dendrimers have been explored and will continue to be developed. Energy transfer efficiency in many dendritic systems is near quantitative, establishing them as viable photosynthetic mimics. The high energy transfer efficiency in these systems has led to their use in optoelectronic devices such as OLEDs and fluorescent sensors. With new easier synthetic approaches to light harvesting dendrimers constantly emerging, it is envisioned that these dendritic systems will be competitive with polymers as organic materials for devices.

References

1. Denti G, Campagna S, Serroni S, Ciano M, Balzani V (1992) J Am Chem Soc 114:2944
2. Serroni S, Denti G, Campagna S, Juris A, Ciano M, Balzani V (1992) Angew Chem Int Ed 31:1493
3. Campagna S, Denti G, Serroni S, Juris A, Venturi M, Ricevuto V, Balzani V (1995) Chem Eur J 1:211
4. Sommovigo M, Denti G, Serroni S, Campagna S, Mingazzini C, Mariotti C, Juris A (2001) Inorg Chem 40:3318
5. Serroni S, Juris A, Venturi M, Campagna S, Resino IR, Denti G, Credi A, Balzani V (1997) J Mater Chem 7:1227
6. Balzani V, Campagna S, Denti G, Juris A, Serroni S, Venturi M (1998) Acc Chem Res 31:26
7. Balzani V, Juris A (2001) Coor Chem Rev 211:97
8. Campagna S, Di Pietro C, Loiseau F, Maubert B, McClenaghan N, Passalacqua R, Puntoriero F, Ricevuto V, Serroni S (2002) Coor Chem Rev 229:67
9. Serroni S, Campagna S, Puntoriero F, Loiseau F, Ricevuto V, Passalacqua R, Galletta M (2003) CR Chimie 6:883
10. Plevoets M, Vögtle F, De Cola L, Balzani V (1999) New J Chem 23:63
11. Zhou X, Tyson DS, Castellano FN (2000) Angew Chem Int Ed 39:4301
12. McClenaghan N, Passalacqua R, Loiseau F, Campagna S, Verheyde B, Hameurlaine A, Dehaen W (2003) J Am Chem Soc 125:5356
13. Vögtle F, Gestermann S, Kauffmann C, Ceroni P, Vicinelli V, De Cola L, Balzani V (1999) J Am Chem Soc 121:12161

14. Balzani V, Ceroni P, Gestermann S, Gorka M, Kauffmann C, Vögtle F (2000) J Chem Soc, Dalton Trans, p 3765
15. Vögtle F, Gorka M, Vicinelli V, Ceroni P, Maestri M, Balzani V (2001) Chem Phys Chem 12:769
16. Vicinelli V, Ceroni P, Maestri M, Balzani V, Gorka M, Vögtle F (2002) J Am Chem Soc 124:6461
17. Kawa M, Fréchet JMJ (1998) Chem Mater 10:286
18. Kawa M, Fréchet JMJ (1998) Thin Solid Films 331:259
19. Pitoisa C, Hult A, Lindgren M (2005) J Lumin 111:265
20. Balzani V, Vögtle F (2003) CR Chimie 6:867
21. Campagna S, Denti G, Sabatino L, Serroni S, Ciano M, Balzani V (1989) Gazz Chim Ital 119:415
22. Campagna S, Denti G, Sabatino L, Serroni S, Ciano M, Balzani V (1989) J Chem Soc, Chem Commun, p 1500
23. Denti G, Serroni S, Campagna S, Ricevuto V, Balzani V (1991) Inorg Chim Acta 182:127
24. Denti G, Serroni S, Campagna S, Ricevuto V, Balzani V (1991) Coord Chem Rev 111:227
25. Campagna S, Denti G, Serroni S, Ciano M, Balzani V (1991) Inorg Chem 30:3728
26. Denti G, Campagna S, Sabatino L, Serroni S, Ciano M, Balzani V (1990) Inorg Chim Acta 176:175
27. Juris A, Balzani V, Barigelletti F, Campagna S, Belser P, von Zelewsky A (1988) Coor Chem Rev 84:85
28. McClenaghan ND, Loiseau F, Puntoriero F, Serroni S, Campagna S (2001) Chem Comm, p 2634
29. Hecht S, Fréchet JMJ (2001) Angew Chem Int Ed 40:74
30. Vögtle F, Plevoets M, Nieger M, Azzellini GC, Credi A, De Cola L, De Marchis V, Venturi M, Balzani V (1999) J Am Chem Soc 121:6290
31. Issberner J, Vögtle F, De Cola L, Balzani V (1997) Chem Eur J 3:706
32. Tyson DS, Luman CR, Castellano FN (2002) Inorg Chem 41:3578
33. Tyler TL, Hanson JE (1999) Chem Mater 11:3452
34. Vögtle F, Gestermann S, Kauffmann C, Ceroni P, Vicinelli V, Balzani V (2000) J Am Chem Soc 122:10398
35. Balzani V, Ceroni P, Gestermann S, Kauffmann C, Gorka M, Vögtle F (2000) Chem Comm, p 853
36. Kido J, Okamoto Y (2002) Chem Rev 102:2357
37. Kuriki K, Koike Y, Okamoto Y (2002) Chem Rev 102:2347
38. Haas Y, Stein G (1971) J Phys Chem 75:3677
39. Stein G, Wurzberg E (1975) J Chem Phys 62:208
40. Xu Z, Moore JS (1993) Angew Chem Int Ed 32:246
41. Xu Z, Moore JS (1993) Angew Chem Int Ed 32:1357
42. Xu Z, Kahr M, Walker KL, Wilkins CL, Moore JS (1994) J Am Chem Soc 116:4537
43. Kopelman R, Shortreed M, Shi Z-Y, Tan W, Xu Z, Moore JS, Bar-Haim A, Klafter J (1997) Phys Rev Lett 78:1239
44. Xu Z, Moore JS (1994) Acta Polym 45:83
45. Devadoss C, Bharathi P, Moore JS (1996) J Am Chem Soc 118:9635
46. Tretiak S, Chernyak V, Makumel S (1998) J Phys Chem B 102:3310
47. Shortreed MR, Swallen SF, Shi Z-Y, Tan W, Xu Z, Devadoss C, Moore JS, Kopelman R (1997) J Phys Chem B 101:6318
48. Swallen SF, Kopelman R, Moore JS, Devadoss C (1999) J Mol Struct 485-486:585

49. Swallen SF, Zhu Z, Moore JS, Kopelman R (2000) J Phys Chem B 104:3988
50. Zhu A, Bharathi P, White JO, Drickamer HG, Moore JS (2001) Macromolecules 34:4606
51. Kleiman VD, Melinger JS, McMorrow D (2001) J Phys Chem B 105:5595
52. Gaab KM, Thompson AL, Xu J, Martinez TJ, Bardeen CJ (2003) J Am Chem Soc 125:9288
53. Gong L-Z, Hu Q-S, Pu L (2001) J Org Chem 66:2358
54. Pugh VJ, Hu Q-S, Zuo X, Lewis FD, Pu L (2001) J Org Chem 66:6136
55. Peng Z, Pan Y, Yu B, Zhang J (2000) J Am Chem Soc 122:6619
56. Melinger JS, Pan Y, Kleiman VD, Peng Z, Davis BL, McMorrow D, Lu M (2002) J Am Chem Soc 124:12002
57. Pan Y, Lu M, Peng Z, Melinger JS (2003) J Org Chem 68:6952
58. Pan Y, Peng Z, Melinger JS (2003) Tetrahedron 59:5495
59. Ranasinghe MI, Hager MW, Gorman CB, III TG (2004) J Phys Chem B 108:8543
60. Ahn TS, Thompson AL, Bharathi P, Müller A, Bardeen CJ (2006) J Phys Chem B 110:19810
61. Burroughes JH, Bradley DDC, Brown AR, Marks RN, Mackay K, Friend RH, Burns PL, Holmes AB (1990) Nature 347:539
62. Kwok CC, Wong MS (2002) Chem Mater 14:3158
63. Halim M, Pillow JNG, Samuel IDW, Burn PL (1999) Adv Mater 11:371
64. Lupton JM, Hemmingway LR, Samuel IDW, Burn PL (2000) J Mater Chem 10:867
65. Pillow JNG, Halim M, Lupton JM, Burn PL, Samuel IDW (1999) Macromolecules 32:5985
66. Jiang D-L, Aida T (1998) J Am Chem Soc 120:10895
67. Kimura M, Shiba T, Muto T, Hanabusa K, Shirai H (1999) Macromolecules 32:8237
68. Harth EM, Hecht S, Helms B, Malmstrom EE, Fréchet JMJ, Hawker CJ (2002) J Am Chem Soc 124:3926
69. Yan X, Goodson T, Imaoka T, Yamamoto K (2005) J Phys Chem B 109:9321
70. Choi M-S, Aida T, Yamazaki T, Yamazaki I (2001) Angew Chem Int Ed 40:3194
71. Choi M-S, Aida T, Yamazaki T, Yamazaki I (2002) Chem Eur J 8:2667
72. del Rosario Benites M, Johnson TE, Weghorn S, Yu L, Rao PD, Diers JR, Yang SI, Kirmaier C, Bocian DF, Holten D, Lindsey JS (2002) J Mater Chem 12:65
73. Maruo N, Uchiyama M, Kato T, Arai T, Akisada H, Nishino N (1999) Chem Comm, p 2057
74. Kato T, Uchiyama M, Maruo N, Arai T, Nishino N (2000) Chem Lett 29:144
75. Yeow EKL, Ghiggino KP, Reek JNH, Crossley MJ, Bosman AW, Schenning APHJ, Meijer EW (2000) J Phys Chem B 104:2596
76. Onitsuka K, Kitajima H, Fujimoto M, Iuchi A, Takei F, Takahashi S (2002) Chem Commun, p 2576
77. Dichtel WR, Hecht S, Fréchet JMJ (2005) Org Lett 7:4451
78. Loiseau F, Campagna S, Hameurlaine A, Dehaen W (2005) J Am Chem Soc 127:11352
79. Xu TH, Lu R, Qiu XP, Liu XL, Xue PC, Tan CH, Bao CY, Zhao YY (2006) Eur J Org Chem, p 4014
80. Jiang D-L, Aida T (1997) Nature 388:454
81. Aida T, Jiang D-L, Yashima E, Okamoto Y (1998) Thin Solid Films 331:254
82. Wakabayaski Y, Tokeshi M, Jiang D-L, Aida T, Kitamori T (1999) J Lumin 83/84:313
83. Hecht S, Vladimirov N, Fréchet JMJ (2001) J Am Chem Soc 123:18
84. Hecht S, Ihre H, Fréchet JMJ (1999) J Am Chem Soc 121:9239
85. Gilat SL, Adronov A, Fréchet JMJ (1999) Angew Chem Int Ed 38:1422

86. Adronov A, Gilat SL, Fréchet JMJ, Ohta K, Neuwahl FVR, Fleming GR (2000) J Am Chem Soc 122:1175
87. Adronov A, Malenfant PRL, Fréchet JMJ (2000) Chem Mater 12:1463
88. Freeman AW, Fréchet JMJ, Koene SC, Thompson ME (1999) Polym Preprints 40:1246
89. Freeman AW, Koene SC, Malenfant PRL, Thompson ME, Fréchet JMJ (2000) J Am Chem Soc 122:12385
90. Furuta P, Brooks J, Thompson ME, Fréchet JMJ (2003) J Am Chem Soc 125:13165
91. Furuta P, Fréchet JMJ (2003) J Am Chem Soc 125:13173
92. Serin JM, Brousmiche DW, Fréchet JMJ (2002) J Am Chem Soc 124:11848
93. Serin JM, Brousmiche DW, Fréchet JMJ (2002) Chem Comm, p 2605
94. Chrisstoffels LAJ, Adronov A, Fréchet JMJ (2000) Angew Chem Int Ed 39:2163
95. Herrman A, Weil T, Sinigersky V, Wiesler U-M, Vosch T, Hofkens J, De Schryver FC, Müllen K (2001) Chem Eur J 7:4844
96. Maus M, De R, Lor M, Weil T, Mitra S, Wiesler U-M, Herrman A, Hofkens J, Vosch T, Müllen K, De Schryver FC (2001) J Am Chem Soc 123:7668
97. Weil T, Wiesler U-M, Herrman A, Bauer R, Hofkens J, De Schryver FC, Müllen K (2001) J Am Chem Soc 123:8101
98. Weil T, Reuther E, Müllen K (2002) Angew Chem Int Ed 41:1900
99. Gronheid R, Hofkens J, Köhn F, Weil T, Reuther E, Müllen K (2002) J Am Chem Soc 124:2418
100. Cotlet M, Gronheid R, Habuchi S, Stefan A, Barbafina A, Müllen K, Hofkens J, De Schryver FC (2003) J Am Chem Soc 125:13609
101. Métivier R, Kulzer F, Weil T, Müllen K, Basché T (2004) J Am Chem Soc 126:14364
102. Cotlet M, Vosch T, Habuchi S, Weil T, Müllen K, Hofkens J, Schryver FD (2005) J Am Chem Soc 127:9760
103. Pan J, Zhu W, Li S, Zeng W, Cao Y, Tian H (2005) Polymer, p 7658
104. Du P, Zhu WH, Xie YQ, Zhao F, Ku CF, Cao Y, Chang CP, Tian H (2004) Macromolecules 37:4387
105. Takahashi M, Morimoto H, Miyake K, Yamashita M, Kawai H, Sei Y, Yamaguchi K (2006) Chem Commun, p 3084
106. Brousmiche DW, Serin JM, Fréchet JMJ, He GS, Lin T-C, Chung SJ, Prasad PN (2003) J Am Chem Soc 125:1448
107. He GS, Lin T-C, Cui Y, Prasad PN, Brousmiche DW, Serin JM, Fréchet JMJ (2003) Optics Lett 28:768
108. Adronov A, Fréchet JMJ, He GS, Kim KS, Chung SJ, Swiatkiewicz J, Prasad PN (2000) Chem Mater 12:2838
109. Oar MA, Dichtel WR, Serin JM, Fréchet JMJ, Rogers JE, Slagle JE, Fleitz PA, Tan L-S, Ohulchanskyy TY, Prasad PN (2006) Chem Mater 18:3682
110. Dichtel WR, Serin JM, Edder C, Frechet JMJ, Matuszewski M, Tan LS, Ohulchanskyy TY, Prasad PN (2004) J Am Chem Soc 126:5380
111. Oar MA, Serin JM, Dichtel WR, Fréchet JMJ, Ohulchanskyy TY, Prasad PN (2005) Chem Mater 17:2267
112. Schenning APHJ, Peeters E, Meijer EW (2000) J Am Chem Soc 122:4489
113. Balzani V, Ceroni P, Gestermann S, Gorka M, Kauffmann C, Maestri M, Vögtle F (2000) Chem Phys Chem 4:224
114. Balzani V, Ceroni P, Gestermann S, Gorka M, Kauffmann C, Vögtle F (2002) Tetrahedron 58:629
115. Hahn U, Gorka M, Vögtle F, Vicinelli V, Ceroni P, Maestri M, Balzani V (2002) Angew Chem Int Ed 41:3595

Adv Polym Sci (2008) 214: 149–203
DOI 10.1007/12_2008_158
© Springer-Verlag Berlin Heidelberg
Published online: 2 July 2008

Excitonically Coupled Oligomers and Dendrimers for Two-Photon Absorption

Chantal Andraud[1] (✉) · Rémy Fortrie[1] · Cyril Barsu[1] · Olivier Stéphan[2] ·
Henry Chermette[1,3] · Patrice L. Baldeck[2] (✉)

[1]Université de Lyon, Laboratoire de Chimie, CNRS – Ecole Normale Supérieure de Lyon,
46 allée d'Italie, 69364 Lyon Cedex 07, France
chantal.andraud@ens-lyon.fr

[2]Laboratoire de Spectrométrie Physique, CNRS UMR 5588, Université Joseph Fourier de
Grenoble, Bât. E, 140 Avenue de la Physique, BP 87, 38402 St-Martin d'Hères Cedex,
France
patrice.baldeck@ujf-grenoble.fr

[3]Chimie Physique Théorique, CNRS UMR 5180, Université Lyon, Bât. Paul Dirac (210),
43 Boulevard du 11 Novembre 1918, 69622 Villeurbanne Cedex, France

Abstract The first part of this review presents the state of the art on two-photon absorbing molecules. Early works concerned the optimization of dipolar push–pull molecules. Recently, the synthesis of linear and branched centrosymmetrical quadrupolar molecules led to higher nonlinearities, but still far from reaching the fundamental limits of molecular two-photon cross-sections.

The second part of this review summarizes our theoretical and experimental results on fluorene-based oligomers, and branched oligomers (V-shape molecules and dendrimers).

They are model molecules to investigate an alternative approach based on spatial assemblies of nonsubstituted π-electron systems that are coupled by dipole–dipole interactions. In all geometries, two-photon absorption cross-sections have superlinear dependencies that depend on coupling energies and oligomer sizes. These results are well rationalized by an excitonic model based on N interacting three-level systems. Analytical expressions for one-photon and two-photon energies and absorption strengths are derived for linear oligomers. An accurate calculation of large exitonic systems is obtained by diagonalizing the Hamiltonian operator on a reduced basis set.

Keywords Branched molecules · Dendrimer · Excitonic coupling · Fluorene · Oligomer · Two-photon absorption

Abbreviations

NLO	Nonlinear optic
TPA	Two-photon absorption
β, γ	First-order, second-order hyperpolarizabilities
γ_{ijkl}	Second-order hyperpolarizability components following i, j, k, l axis
E	Electric field
ω	Laser frequency
λ_{abs}	Linear absorption wavelength
λ_{TPA}	Two-photon absorption wavelength
σ_{TPA}	Two-photon absorption cross-section
$\chi^{(3)}$	Macroscopic third polarizability or second hyperpolarizability
$\vec{u}$	Unitary vector of space
d	Intermonomeric distance
$\vec{\mu}$	Electric dipole moment operator
$\lvert i \rangle$	Electronic eigenstate i
$\lvert i_1 \ldots i_N \rangle$	Element of the oligomeric or dendrimeric basis set.
$\{\lvert i_1 \ldots i_N \rangle\}$	Oligomeric or dendrimeric basis set
E_i	Energy of $\lvert i \rangle$
e_i or E_{0i}	$E_i - E_0$
ω_{ij}	$(E_i - E_0)/\hbar$
$\vec{\mu}_{ij}$	$\langle i \lvert \vec{\mu}_i \rvert j \rangle$
μ_{ij}	$\lvert \vec{\mu}_{ij} \rvert$
$\Delta\mu_{ij}$	$\lvert \vec{\mu}_i - \vec{\mu}_j \rvert$
$\tau_{i,i+1}^{j,j+1}$	Junction parameter within the excitonic model
V_{ij}	Excitonic perturbation matrix element
f_{ij}	Transition oscillator strength
f_{ij}^{TPA}	Two-photon absorption transition strength
DEANST	N,N-Diethyl-4-(2-nitroethenyl)phenylamine
Nn	Oligomer consisting of N monomers
Vn	V-shape molecule with N monomers on branches
$DnGg$	g-generation dendrimer with N monomers on branches
F–F	fluorene–fluorene junction
F–P–F	fluorene–phenyl–fluorene junction in Vn (V-type) and $DnGg$ (D-type)
L	Metric oligomeric length
α	Power of L in the dependency of σ_{TPA} as a function of L
N (or n)	Number of monomeric units in an oligomer or dendrimer

1
Introduction

Since its theoretical prediction by Göppert-Mayer in 1931 [1], the nonlinear optical (NLO) process of two-photon absorption (TPA) has received considerable attention, owing to numerous relevant applications in several fields [2], such as optical limiting [3–5], three-dimensional microfabrication [7–15], up-converted lasering [16–18], photodynamic therapy [19–22], data storage [23–28], and biomedical imaging [29–33].

Because of this wide range of applications, much effort was dedicated to the design and synthesis of new molecules with optimized TPA efficiency; in this context, the characteristics of the designed molecules (linear absorption, solubility, substituents...) will depend on the targeted application. The TPA response of molecules can be understood in the context of its TPA cross-section σ_{TPA}, which can be measured using different techniques, such as nonlinear transmission, two-photon induced fluorescence and the Z-scan method[1]; although in a pure TPA process σ_{TPA} does not depend on the laser pulse duration, the nonlinear absorption can be more efficient in the nanosecond regime than in the femtosecond one, due to excited state reabsorption phenomena [34].

A NLO active molecule submitted to an electric field E is described by the induced electronic dipole moment μ following the relationship Eq. 1, in which μ_0 is the permanent dipole moment, α is the linear polarization, while β and γ are respectively the first and second hyperpolarizabilities.

$$\mu = \mu_0 + \alpha E + \beta E^2 + \gamma E^3 + \dots \tag{1}$$

σ_{TPA} is related to the imaginary part of the second hyperpolarizability γ following the relationship Eq. 2 [35], in which $\hbar$ is the Planck constant, n the refraction index of the medium and f the local field factor; ω is the incident laser frequency.

$$\sigma_{TPA} = \frac{8\pi^2 \hbar \omega^2}{n^2 c^2} f^4 \mathrm{Im}\, \gamma(-\omega; \omega, \omega, -\omega). \tag{2}$$

Since sTPA is a third-order NLO process, there are no restrictions on the molecular process to occur. The optimization of TPA properties of organic molecules is related to the optimization of the second hyperpolarizability γ, which is given by the Orr and Ward sum-over states (SOS) relationship Eq. 3 [36]. In this expression, $P(i, j, k, l; -\omega; \omega, \omega, -\omega)$ is a permutation operator, $|0\rangle$ is the ground state, while $|m\rangle$, $|n\rangle$ and $|p\rangle$ are excited states. $\langle 0|\mu_i|m\rangle$ and $\langle m|\bar{\mu}_i|n\rangle$ correspond respectively to the components μ^i_{0m} and μ^i_{mn} of the transition dipole moment between $|0\rangle$ and $|m\rangle$ and between $|m\rangle$ and $|n\rangle$ along the axis

---

[1] σ_{TPA} is generally expressed in Göppert-Mayer (GM) with $1\,GM = 10^{-50}\,cm^4/$ photonmolecule.

i of the molecule; $\langle n|\bar{\mu}_i|n\rangle$ is the component of the static dipole moment difference $\mu_{nn}^i = \Delta\mu_{0n}^i$ along the axis i. $\hbar\omega_{m0}$ represents the energy E_{0m} of the excited state $|m\rangle$, while Γ_{m0} is the homogeneous width associated to the state $|m\rangle$.

$$\gamma_{ijkl}(-\omega; \omega, \omega, -\omega) = \frac{1}{6\hbar^3} P(i, j, k, l; -\omega; \omega, \omega, -\omega)$$

$$\times \left[\sum_{m\neq 0} \sum_{n\neq 0} \sum_{p\neq 0} \frac{\langle 0|\mu_i|m\rangle\langle m|\bar{\mu}_j|n\rangle\langle n|\bar{\mu}_k|p\rangle\langle p|\mu_l|0\rangle}{(\omega_{m0}-\omega-i\Gamma_{m0})(\omega_{n0}-2\omega-i\Gamma_{n0})(\omega_{p0}-\omega-i\Gamma_{p0})} \right.$$
$$\left. - \sum_{m\neq 0} \sum_{n\neq 0} \frac{\langle 0|\mu_i|m\rangle\langle m|\mu_j|0\rangle\langle 0|\mu_k|n\rangle\langle n|\mu_l|0\rangle}{(\omega_{m0}-\omega-i\Gamma_{m0})(\omega_{n0}-\omega-i\Gamma_{n0})(\omega_{n0}+\omega-i\Gamma_{n0})} \right]. \tag{3}$$

In order to compare theoretical σ_{TPA} values to those measured in solution, the orientationally averaged value of γ is generally calculated following Eq. 4:

$$\langle\gamma\rangle = \frac{1}{15} \sum_{i,j} (\gamma_{iijj} + \gamma_{ijij} + \gamma_{ijji}) \quad i, j = x, y, z. \tag{4}$$

Different parameters of Eq. 3, such as excited-state energies and dipole moments, must be tuned to obtain high TPA cross-sections. This has been obtained using different molecular approaches optimizing the second-order hyperpolarizability γ. We will present in the first part, a review of the most efficient molecules for TPA, following their symmetry. Although a wide priority will be given to branched systems (octupolar and dendritic molecules) developed in order to optimize TPA properties by dipolar interactions, linear systems (dipolar and quadrupolar) with high TPA cross-section values will be also described.

2
State of the Art on Two-Photon Absorbing Molecules

2.1
Linear Molecules

Linear organic molecules were firstly considered for the design of TPA efficient molecules. This family can be divided into two groups (Fig. 1): (1) *dipolar molecules* constituted by a noncentrosymmetric structure, with a linear delocalized electrons system substituted at each end by electron donor D and acceptor groups A (push–pull molecules); (2) *quadrupolar systems*, consisting of symmetric molecules with a π electrons system substituted by donor or acceptor groups.

In this case, the second-order hyperpolarizability γ can be considered as a unidimensional parameter defined along the axis x of the molecule following the single component γ_{xxxx}. Equation 3 can be simplified into the relationship Eq. 5, when applying the three-level model in which the lowest excited state $|1\rangle$ and the two-photon excited state $|2\rangle$ are considered to be the most significant

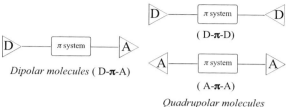

Fig. 1 Schematic representation of dipolar and quadrupolar molecules (A and D represents, respectively, electron acceptor and donor)

contributors to the value of γ_{xxxx} [37]. This model is illustrated in Fig. 2.

$$
\gamma_{xxxx} \propto
\begin{bmatrix}
- \dfrac{\mu_{01}^4}{(E_{01}-\hbar\omega-i\Gamma)^2(E_{01}+\hbar\omega-i\Gamma)}\,|N \\[2ex]
+ \dfrac{\mu_{01}^2 \Delta\mu_{01}^2}{(E_{01}-\hbar\omega-i\Gamma)^2(E_{01}-2\hbar\omega-i\Gamma)}\,|D \\[2ex]
+ \dfrac{\mu_{01}^2 \mu_{12}^2}{(E_{01}-\hbar\omega-i\Gamma)^2(E_{02}-2\hbar\omega-i\Gamma)}\,|T
\end{bmatrix}
\tag{5}
$$

This expression consists of three terms: (1) the negative term N, which contributes only for a one-photon resonance and can be neglected in the case of TPA; (2) the two-photon term T, which is related to a two-photon resonance with the excited state $|2\rangle$ and takes the specific two-photon excited states participation into account (case a in Fig. 2); (3) the dipolar term D, which corresponds to a two-photon resonance with the low-lying excited state $|1\rangle$ (as illustrated by configuration b in Fig. 2) and is zero for centrosymmetric molecules.

This model has been examined for numerous organic molecules; an example is shown in Fig. 3, in which variations of σ_{TPA} values are plotted as a function of the number of excited states involved in the SOS rela-

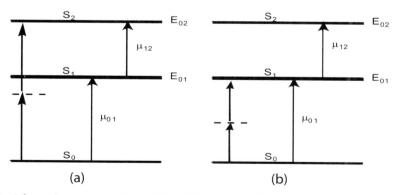

Fig. 2 Schematic representation of the TPA process. (**a**) In the case of the three-level model for symmetric systems; (**b**) in the case of the two-level model for noncentrosymmetric systems

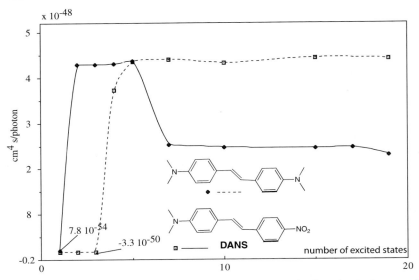

Fig. 3 σ_{TPA} variations with the number of excited states involved in the SOS expression (3) for the 4-dimethylamino-4-nitrostilbene and the dimethylamino)-E-stilbene

tionship Eq. 3 for a dipolar molecule (4-dimethylamino-4-nitrostilbene) and a symmetric one (dimethylamino)-E-stilbene [38]. This figure shows that the convergence for γ is rapidly reached for seven excited states and that only lowest excited states contribute mainly to σ_{TPA} values.

2.1.1
Push–Pull Molecules

This type of molecule is usually noted as D–π–A, in which D and A are respectively electron donor and acceptor groups and π the central conjugated system (Fig. 1). This noncentrosymmetric structure has been widely studied for its high efficiency in quadratic NLO, quantified at the molecular level by the first hyperpolarizability β in relationship Eq. 1. This coefficient can be expressed following the relationship Eq. 6, assuming the validity of the two-level model [39] presented in Fig. 2b; in the case of push–pull molecules, this two-level model is valid for TPA at E_{01} resonance, while the three-level model must be considered at E_{02} resonance. In the former case, the same parameters μ_{01}, $\Delta\mu_{01}$ and E_{01} are involved in both expressions of β_{xxx} and γ_{xxxx} (Eqs. 6 and 5, respectively); this implies that molecules, designed for high β values, should present also high γ values at least at $E_{01} = \hbar\omega_{01}$ resonance.

$$\beta_{xxx} = \frac{3(\mu_{01})^2 \Delta\mu_{01}}{2(\hbar\omega_{01})^2} \frac{\omega_1^4}{(\omega_{01}^4 - 4\omega^2)(\omega_{01}^2 - \omega^2)} \cdot \tag{6}$$

Much theoretical as well experimental studies have been focussed on the design of efficient TPA molecules [40–51]. As for quadratic NLO, the best trade-off between transparency and TPA response is obtained by tuning the conjugation of the charge transfer (CT) system and the strength of the substituents A and D of the molecule; this leads to the shift of excited states. This trend is illustrated in Fig. 4 by the work of Reinhardt et al. [52], from which several structure/TPA efficiency relationships enhancing σ_{TPA} values could be deduced: (1) the role of the conjugated system can be observed in the conjugation extension from **R1** to **R2**, in the planarity increase from **R3** to **R4**, in the substitution of the naphtyl by the biphenyl from **R2** to **R3**, and in the extension of the fluorene to the bifluorene (**R4** to **R5**); (2) the role of the acceptor by the substitution of the 4-pyridyl by the 2-pyridyl (**R4** and **R10**, respectively); (3) the role of the fluorene moieties pendant alkyl chains on the molecular aggregation; (4) the role of the donor in **R4** and **R6**; (5) the role of hydrogen bonds in **R7** with respect to **R6**.

Nunzi et al. [53] showed the two-level model was well adapted for the 4-(N-(2-hydroxyethyl-N-ethyl)-amino-4′-nitrobenzene (DR1) dye, while a three-level model had to be considered for the 4-dibutylamino-4′-nitrobenzene (DBANS) system, due to electron–vibration interactions; values of μ_{01}, μ_{12} and $\Delta\mu_{01}$ could be deduced for both molecules.

Brédas et al. studied theoretical TPA properties of the 4-dimethylamino-4′-formylstilbene (DAFS) with the degree of π bond-order alternation (BOA) by the application of an external field [54]. This work shows two possible strategies to optimize TPA response in stilbenic molecules, as shown in Fig. 5: (1) In the case of TPA resonance in the lowest excited state at E_{01} energy, the D term of Eq. 5 presents the main contribution and σ_{TPA} varies in the same way as β and $\Delta\mu_{01}$ as a function of BOA, with a maximum for an intermediate regime between the neutral **Br1** and the cyanine **Br2** limits. (2) For a TPA resonance in the specific two-photon state S_2, the structure presents huge σ_{TPA} values for the cyanine structure near the double resonance at $E_{01} = \frac{E_{02}}{2}$.

Barzoukas and Blanchard-Desce proposed an approach of molecular engineering using multivalence-bond state models [55]. Push–pull polyenes were shown also to present an enhancement of the TPA response and a loss of transparency of molecules, as a function of the increase of the polyenic chain length [56, 57]. Trends observed in these polyenic systems are supported by the large third-order optical nonlinearities measured in asymmetric carotenoids, in which the role of the large value of dipole moment difference $\Delta\mu$ was shown [58].

In good agreement with the model described in Fig. 2b for push–pull molecules, TPA spectra match one-photon absorption curves; this confirms that the value of σ_{TPA} is dominated by the dipolar term of Eq. 5 and that dipolar molecules efficient in quadratic NLO processes (see Eq. 6) should present high TPA responses. This trend has been illustrated by work on azo-aromatic

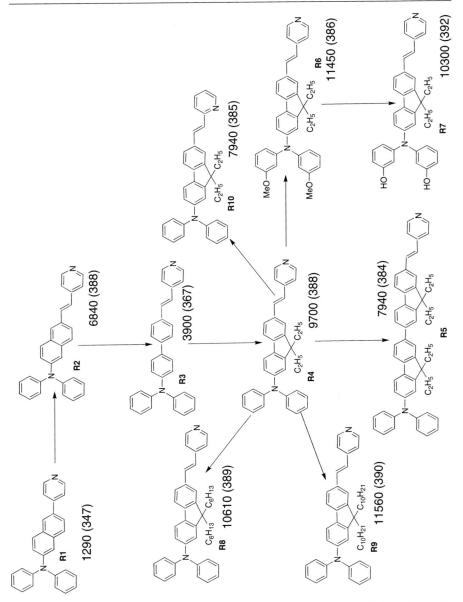

Fig. 4 Example of the relationship between molecular structure/TPA efficiency in push–pull molecules containing pyridyl and biphenyl amino moieties; $\sigma_{TPA}(\lambda_{max})$ in THF is expressed in GM for a laser excitation at 800 nm, and the linear absorption λ_{max} is in nm

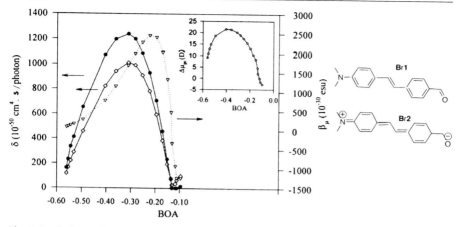

Fig. 5 Variations of $\delta = \sigma_{TPA}$ in DAFS with BOA: fully SOS values ($\diamond$), contribution of the *D* term in Eq. 5 ($\bullet$); β values ($\triangledown$); *insert*: variations of $\Delta\mu_{ge} = \Delta\mu_{01}$

molecules, in which a TPA resonance in the one-photon state $|1\rangle$ with a more efficient peak at resonance with the higher two-photon excited state $|2\rangle$ [59].

Belfield et al. studied different dipolar molecules based on fluorene derivatives for the π-system with a diphenylamine and a benzothiazole as the electron donor and acceptor, respectively (Fig. 6).

Variations of σ_{TPA} and of the intermediate state resonance enhancement ISRE, the ratio between the nondegenerate TPA cross-section and the degen-

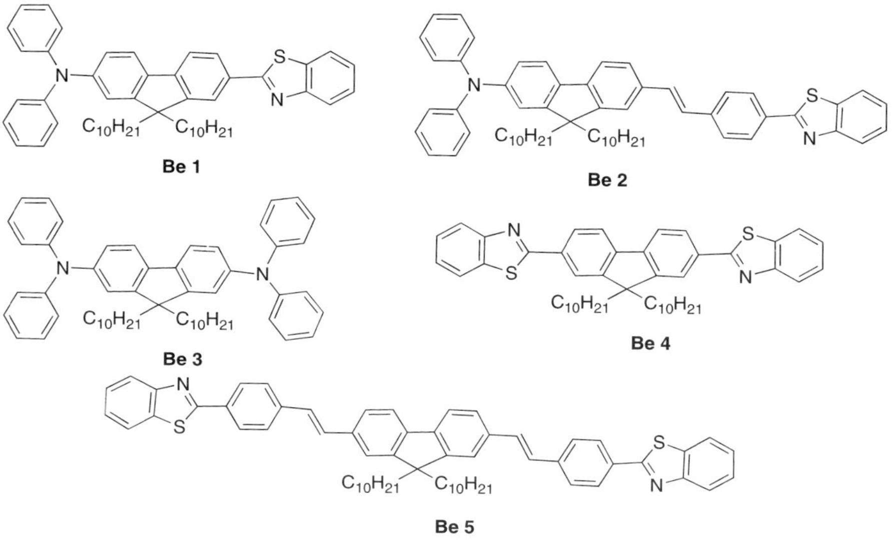

Be 1

Be 2

Be 3

Be 4

Be 5

Fig. 6 Benzothiazolyl fluorene derivatives **Be1–Be5**

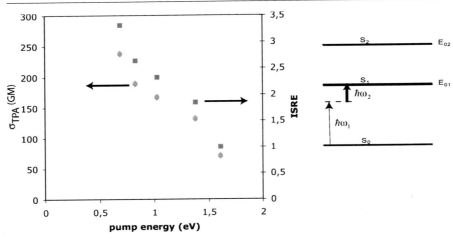

Fig. 7 Variations of σ_{TPA} and of ISRE in **Be1**, as a function of the pump energy $\hbar\omega_2$ in a nondegenerate TPA process

erate one (that corresponds to $E_{01} = \hbar\omega_1 + \hbar\omega_2$ in the nondegenerate TPA process, $\hbar\omega_1$ being the probe energy) are plotted in Fig. 7 for **Be1** as a function of $\hbar\omega_2$; the possibility of a three-fold enhancement of σ_{TPA} with respect to data observed in a degenerate TPA process has been shown by tuning the laser pump energy $\hbar\omega_2$. Furthermore, an enhancement of 20-times the response in the case of the degenerate process was observed for higher excited states [60]. This last trend is confirmed in the case of **Be2** with a more extended π-system, for which σ_{TPA} values of 1530 and 525 GM have been measured at 610 and 790 nm, respectively, in linear absorption bands [61].

The two-photon excited fluorescence properties of compounds **Fg1–9**, bearing B(Mes)$_2$ as acceptor, with different donors and π-systems, have been studied by Fang et al. (Fig. 8) [62]. From linear absorption, the strength of the B(Mes)$_2$ group has been found to be intermediate between those of CN and NO$_2$ and higher than that of the benzothiazole (**Fg10**); this group has been shown to induce significant charge transfer in the ground state and more predominantly in the first excited state. The molecule **Fg3** has the highest σ_{TPA} value at 800 nm; an efficiency of 300 GM, which is of an order-of-magnitude higher than that of the fluorescein, with a fluorescence quantum yield of 0.9. These observations provide evidence for continued interest of this class of molecules for TPA applications.

The group of Abbotto observed a tunable two-photon pumped lasing emission in the range of 500–600 nm with the same laser pumping at 790 nm, by the modulation of the chemical structure of heterocycle-based derivatives (Table 1) [63].

A series of diphenylaminofluorene derivatives (Fig. 9) was studied in order to reach higher TPA responses than that of **AF-50**, a TPA bench-

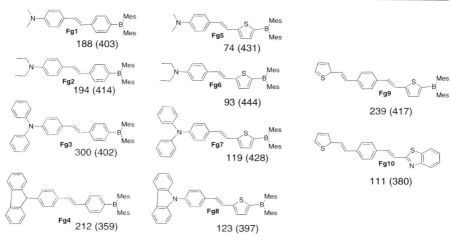

Fig. 8 Trivalent boron derivatives **Fg1–Fg8**; σ_{TPA} (λ_{max}) is expressed in GM for a laser excitation at $2\lambda_{max}$, and the linear absorption λ_{max} is in nm

Table 1 Laser emission wavelengths obtained by two-photon pumping at 790 nm for different molecules

Molecules	λ_{em} (nm)
	522
	532
	555
	557
	576

mark in the nanosecond regime with improved thermal and photochemical stabilities [64]; for this purpose, 2-benzothiazolyl, benzoyl, 2-benzoxazolyl, 2-quinoxalinyl, 2-(N-phenylbenzoimidazolyl) and 2-(4,5-diphenylimidazolyl) groups have been successively introduced as acceptors in **AF-240**, **AF-370**, **AF-390**, **AF-260**, **AF-386**, **AF-385**, respectively. The conclusion was that the 2-benzothiazolyl and the benzoyl derivatives **AF-240** and **AF-370** respectively were good candidates to replace **AF-50** in nanosecond TPA applications.

Dithienothiophene derivatives **DTT1** and **DTT2** (Fig. 10), with a 2-phenyl-5-(4-*tert*-butyl)-1,3,4-oxadiazole as acceptor and respectively with carbazole and triphenylamine as donors, have been also compared to **AF-50** [65]; the comparison between **DTT1** and **DTT2** demonstrated the higher efficiency of the triphenylamine with respect to that of the *N*-ethylcarbazole. On the other hand, **DTT2** has been found to be 6-times more efficient than **AF-50** (measured with a σ_{TPA} value of 19 400 GM in benzene)[2]; this study contributed to show a great interest in dithienothiophene as a transmitting bridge (for the comparison, acceptor groups pyridine of **AF-50** and oxadiazole of **DTT2** have been considered to present similar efficiencies).

Recently, dyes based on the dicarboxamide pyridine as acceptor have been designed to obtain a high two-photon excited fluorescence and second harmonic signals for potential membrane measurement (Table 2) [66]. These systems have been found to present σ_{TPA} values close to those of the most ef-

---

[2] It is worth to note that, in the nanosecond regime, several values of σ_{TPA} for **AF-50** were published: 9700 and 11 560 GM in THF at 800 nm (see molecules **R4** and **AF-50** in Figs. 4 and 9 respectively and 19 400 GM in benzene in [65]).

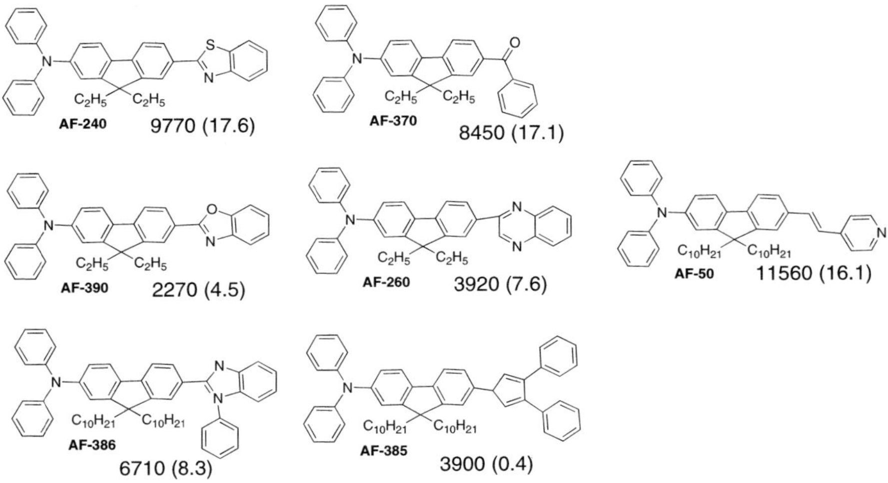

AF-240   9770 (17.6)

AF-370   8450 (17.1)

AF-390   2270 (4.5)

AF-260   3920 (7.6)

AF-50   11560 (16.1)

AF-386   6710 (8.3)

AF-385   3900 (0.4)

Fig. 9 Diphenylaminofluorene derivatives AF; σ_{TPA} at 800 nm in THF is expressed in GM (GM g^{-1}), and the molecular weight is in g

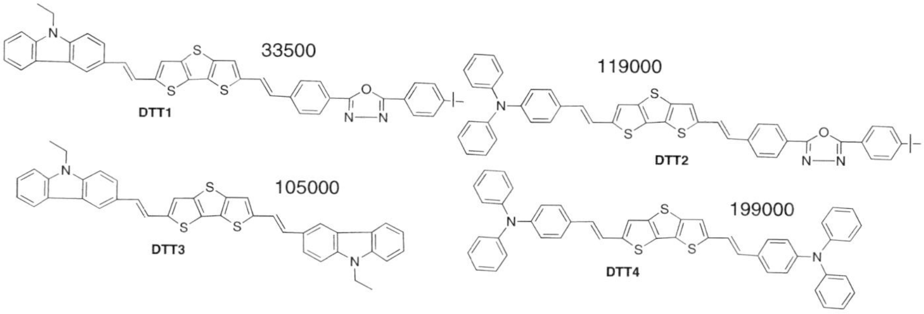

Fig. 10 Dithienothiophene derivatives **DTT1–DTT4**; σ_{TPA} measured in 1,1,2,2-Tetrachloro-ethane at 810 nm is expressed in GM

ficient systems used in two-photon microscopy. These ligands were also used to induce efficient TPA antenna effects for lanthanide fluorescence [67].

It is important to note that molecules with TPA in the NIR spectral region have been less than those in the visible. Dipolar systems of $D-\pi-D'-\pi-A'-\pi-A$ type, bearing intermediate pyrrole and thiazole as donor and acceptor groups (Fig. 11), were studied at telecommunications wavelengths by Marder et al. [68]. These chromophores were shown to exhibit high nondegenerate σ_{TPA} values of 1500 GM between 1.3 and 1.55 μm.

Our group in Lyon developed heptamethine dyes with promising TPA cross sections in the range 1400–1600 nm [69]. **An1** (Fig. 12) exhibits a σ_{TPA} value of 792 GM at 1445 nm; these molecules, have many promising properties in terms of synthesis, stability and solubility and may have potential for practical applications in biology and optical limiting.

Table 2 Dicarboxamide pyridine-based chromophores

Molecules	λ_{TPA}^{max} (nm)	σ_{TPA} (GM)	$\mu\beta_0$ (10^{-48} esu)
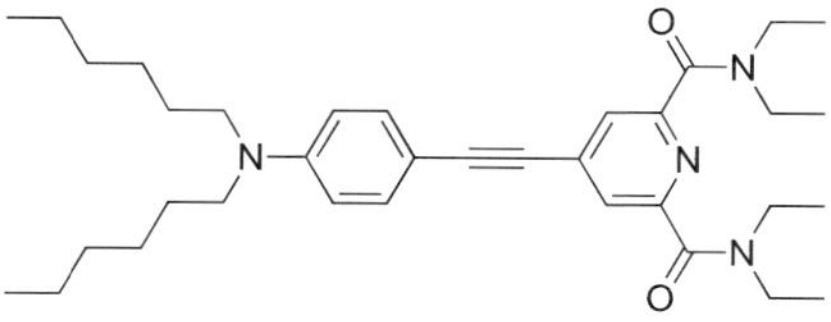	770	780	177
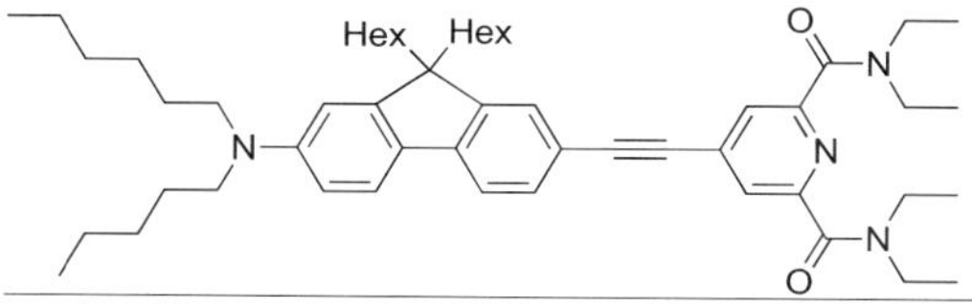	830	1146	249

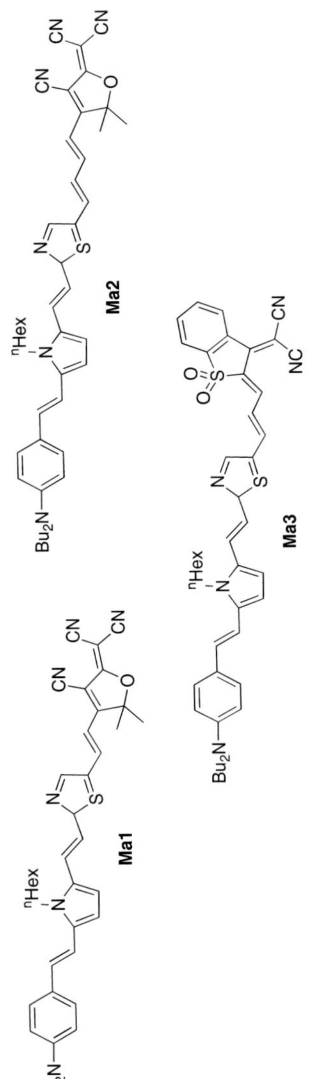

Fig. 11 Dipolar systems of **Ma1–Ma3** for TPA at telecommunications wavelengths

Although asymmetric dipolar systems have been shown to exhibit high σ_{TPA} values at a variety of ranges of wavelengths, the linear absorption of these molecules is generally red-shifted with respect to their symmetrical analogs. Similarly, due to the main contribution of the dipolar term in Eq. 5 for the lowest TPA allowed state, the TPA transparency in push–pull

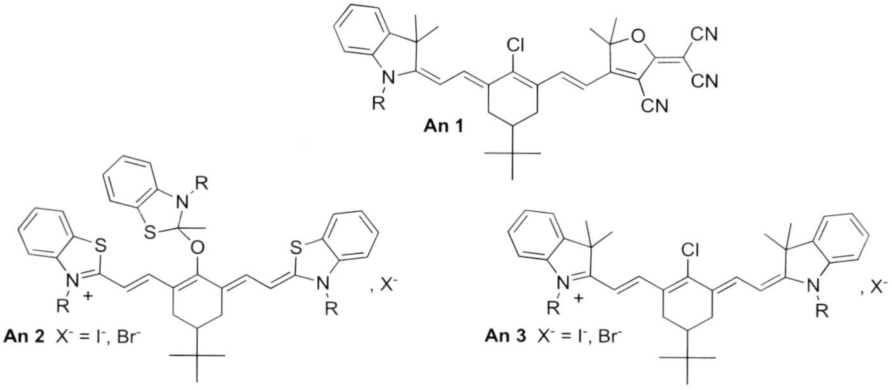

Fig. 12 Heptamethine dyes **An1–An3** for TPA in the NIR

molecules is lower; the main drawback of this trend for practical applications is the overlap between linear and TPA spectra, as has been discussed in [14].

2.1.2
Quadrupolar Molecules

The structure of quadrupolar molecules is generally noted as $D-\pi-D$ or $A-\pi-A$ (see Fig. 1). For these molecules, the term $\Delta\mu_{01}$ of the relationship Eq. 5 cancels and this equation can be rewritten as Eq. 7, and molecules fulfill the three-level model represented in Fig. 2a.

$$\sigma_{\text{TPA}} \infty \frac{\mu_{01}^2 \mu_{12}^2}{(E_{01} - E_{02}/2)^2 \Gamma} \, . \tag{7}$$

The pioneering work of Brédas, Marder, Perry et al. in this field appeared in 1998 [70]. From the *trans*-stilbene **Q1** as starting molecule, the authors designed quadrupoles based on a symmetrical CT from the ends to the molecule center (**Q2-6**) or vice-versa (**Q7-9**), following different motifs : $D-\pi-D$ for **Q1-3**, $D-\pi-D'-\pi-D$ for **Q4** and **Q5**, $D-\pi-A-\pi-D$ for **Q6**, $A-\pi-D-\pi-A$ for **Q7-9** (Table 3). These types of structures were shown to be very efficient, with σ_{TPA} values up to 4400 GM for **Q9**. Quantum chemical calculations pointed to a large electronic redistribution with excitation, which correlated with a significant increase of μ_{12}. This is the main factor in the σ_{TPA} enhancement between **Q1** and **Q2**, for which the terminal substitution leads to a red-shift of excited states energies. The intermediate substitution by donors in **Q4** does not modify spectroscopic properties with respect to those of **Q3** without intermediate donor; an inverse trend is observed for **Q6**. The conjugation extent (**Q2** vs. **Q3**, **Q4** vs. **Q5**) induces variations of parameters in Eq. 7. The highest value of σ_{TPA} has been observed in **Q9**, which bears a strong acceptor; this system exhibits also the strongest red-shifted TPA properties.

Molecules	λ_{abs}^{exp} (nm)	$\lambda_{TPA}^{exp}\left[\lambda_{TPA}^{th}\right]$ (nm)	$\sigma_{TPA}^{exp}\left[\sigma_{TPA}^{th}\right]$ (GM)	μ_{01} (D)	$\Delta\mu_{01}$ (D)
Q1		514 [466]	124 [27]	7.1	3.1
Q2	374	605 [529]	210 [202]	8.8	7.2
Q3	408	730 [595]	995 [681]	13.3	9.1
Q4	428	730 [599]	900 [670]	13.1	9.3
Q5	456	775 [620]	1250 [713]	14.6	6.0
Q6	472	835 [625]	1940 [950]	12.4	11.9
Q7	513	825 [666]	480 [570]	14.3	8.3
Q8	554	970 [–]	1750 [–]	–	–
Q9	618	975 [–]	4400 [–]	–	–

Table 3 Properties of quadrupolar bis(styryl)benzene chromophores

A similar study from the same group confirmed the strong influence of donors on excited states properties and particularly on μ_{12}, while μ_{01} and the detuning energy are correlated to the conjugation extent of molecules [71].

This work led to many molecular engineering investigations on quadrupolar molecules for TPA [72–88]. Quadrupolar polyenes have been theoretically studied by Cho et al. [89]. Parameters involved in Eq. 7 have been investigated: in general increasing the strength of donor or acceptor induces an increase of the product $\mu_{01}\mu_{12}$ in the numerator, and simultaneously a decrease of the detuning energy term of the denominator; a saturation in the σ_{TPA} enhancement has been obtained for about 14 double bonds.

In addition to general work based on engineering for a molecular structure/TPA relationship building, more specific studies targeting different TPA-related applications have been performed; in this regard molecules, fulfilling requirements for the selected application, have been designed. Because of the need for TPA-based optical limitors in the visible region (between 450–650 nm) with a good transparency, triarylamine derivatives displayed in Fig. 13 have been synthesized [90]. For the purpose of optical limiting in the visible region, specific requirements for molecules were high solubility (of the order of one mol L^{-1}) and a linear absorption cut-off at 420 nm. While the above relationship structure/TPA properties in molecules **Q1–9** was based on the optimization of the charge transfer at the expense of the molecule transparency, the visible transparency condition for optical limiting requires the optimization of the trade-off between TPA properties and the linear absorption of molecules. This was realized by molecular engineering around the moderately conjugated biphenyl **RA1**; the optimization of the charge transfer by the increase of the planarity, in **RA2** and **RA3** or **RA4** and **RA5**, was considered. All molecules **RA1–5** exhibit a $\lambda_{\text{cut-off}} \leq 420$ nm. Broad band TPA spectra were obtained in the visible region; the planar conjugated fluorene **RA2** was shown to exhibit the best TPA efficiency. Besides the molecular engineering of the optimization of σ_{TPA}, molecular engineering for the excited states dynamics was reported for this family [91].

Recently, the group of Marder obtained high TPA cross-sections of 1000–5000 GM in the range 600–650 nm using pyrrole or dialkoxythiophene bridges in D–π–D'–π–D type molecules **Ma4** and **Ma5** (Fig. 14) [92]. Quantum chemical calculations have ascribed this large TPA response to the role of high energies of two-photon excited states with a large detuning energy term and an unusually large transition dipole moment μ_{12} (Eq. 7).

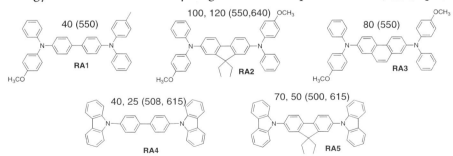

Fig. 13 Triarylamine derivatives **RA1–RA5** for TPA between 450 and 650 nm; $\sigma_{TPA}(\lambda_{TPA})$ is expressed in GM and the TPA wavelength λ_{TPA} is in nm

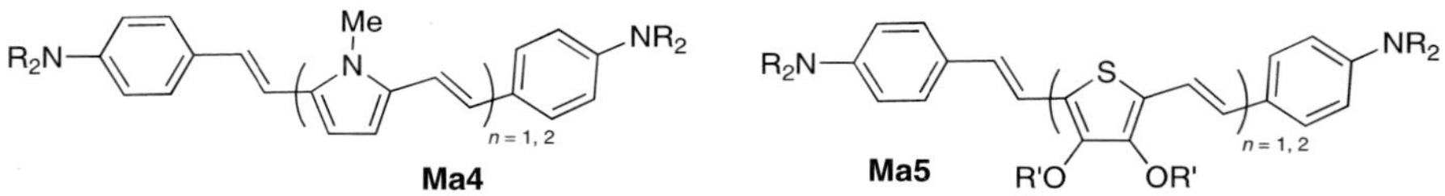

Fig. 14 Symmetric systems **Ma4** and **Ma5** for TPA in the visible

On the contrary, D–π–A–π–D molecules are generally known to provide high σ_{TPA} values from the lowest two-photon excited states with a weaker detuning energy term; this has been observed recently in conjugated porphyrins [93], in extended phenylenevinylene oligomers for which σ_{TPA} values up to 5300 GM have been obtained [94], and in squaraines [95].

Many molecules were designed for TPA properties in the range of wavelengths used in two-photon microscopy between 750 and 950 nm. Ferrocene-based chromophores have been shown to lead to interesting TPA properties; their efficiency depends strongly on the conjugation length and on the strength of the acceptor. Quadrupolar molecules have been shown to be 4-times more efficient than their dipolar analogs [96]. TPA properties of distyrylbenzenes and polyenes have been studied in the spectral range of 710–960 nm [57]. D–π–A–π–D type molecules are shown to present the largest efficiencies, with $A = CN$ or SO_2R; as expected for symmetric molecules, two-photon excited states have been found to be energetically higher than the lowest singlet state. The fluorene derivative **Be3** (Fig. 6) has been studied by nondegenerate two-photon absorption [60]; a efficiency 5-times higher than that of the dipolar analog **Be1** has been obtained. TPA properties of quadrupoles **Be4** and **Be5** have been compared to those of the dipolar **Be2** (Fig. 6) by the same group [61]. TPA spectra consist of two bands at ~ 800 and 600 nm, with a higher intensity for that at 600 nm, which corresponds to TPA into the $|S_2\rangle$ state; the conjugation increase from **Be4** to **Be5** led to an increase of σ_{TPA} values from 370 to 6000 GM at 600 nm, while the push–pull molecule **Be2** exhibits a response 4-times weaker at the same wavelength. The same trend has been observed for dithienothiophene derivatives **DTT3** and **DTT4** when comparing their efficiencies to those of dipolar molecules **DTT1** and **DTT2**, respectively (Fig. 10) [65]; moreover in good agreement with the observations described above from comparison between **DTT1** and **DTT2**, the higher σ_{TPA} value obtained for **DTT4** (199 000 GM[3]) compared to that of **DTT3** (105 000 GM) also points to the stronger donor efficiency of the triphenylamine with respect to that of N-ethylcarbazole. The pseudoconjugated cyclodiborazane core has been shown by Nicoud et al. to induce strong electronic interactions between both parts of the molecule and to enhance

---

[3] It is worth noting that a σ_{TPA} value of 3 orders of magnitude lower was measured for **DTT4** in the femto second regime; this discrepency was ascribed to excited state reabsorption when using longer laser pulses.

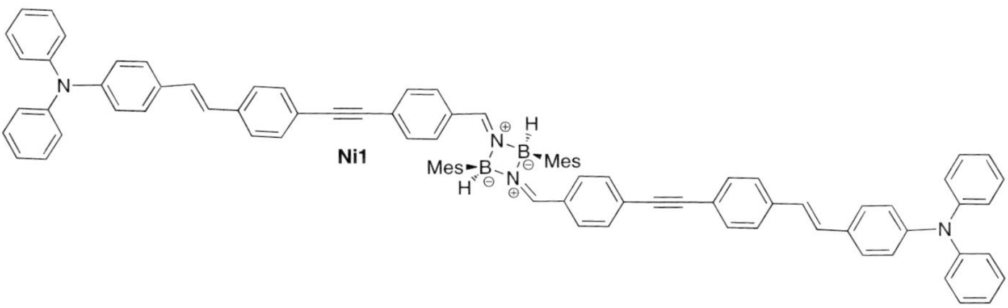

Fig. 15 Cyclodiborazane core chromophore **Ni1**

strongly TPA properties [97]. A TPA cross-section of 1350 GM was found at 830 nm for molecule **Ni1** in Fig. 15.

The group of Blanchard-Desce synthesized a family of molecules based on the general scheme displayed in Fig. 16, in order to deduce wide and comprehensive relationships in structure/physical properties (TPA and fluorescence) in the range 700–900 nm for two-photon excited fluorescence-based applications [98, 99]. The influence of the conjugated core, of the length or the nature of linkers on TPA and spectroscopic properties was systematically studied. A TPA cross-section of 4200 GM at 710 nm has been measured, with high two-photon fluorescence properties.

TPA photoinitiators, with good reducing properties for radical creation, have been designed for 3D microfabrication following incident laser wavelengths. **RA1** and **RA2** (Fig. 13) have been used at 532 nm [100, 101]. For the

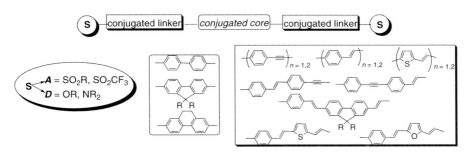

Fig. 16 Two-photon excited fluorescent push–push and pull–pull systems

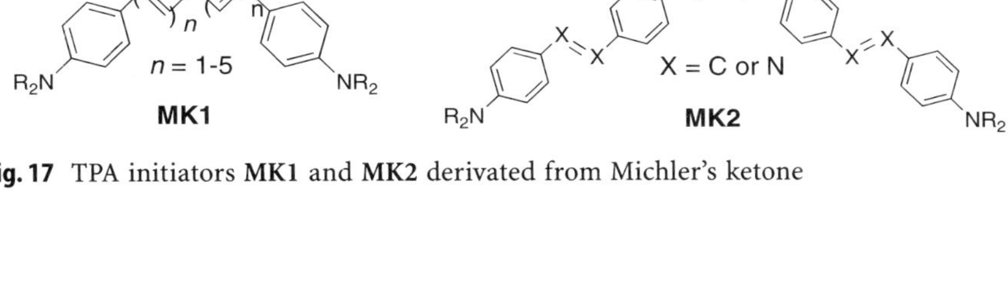

Fig. 17 TPA initiators **MK1** and **MK2** derivated from Michler's ketone

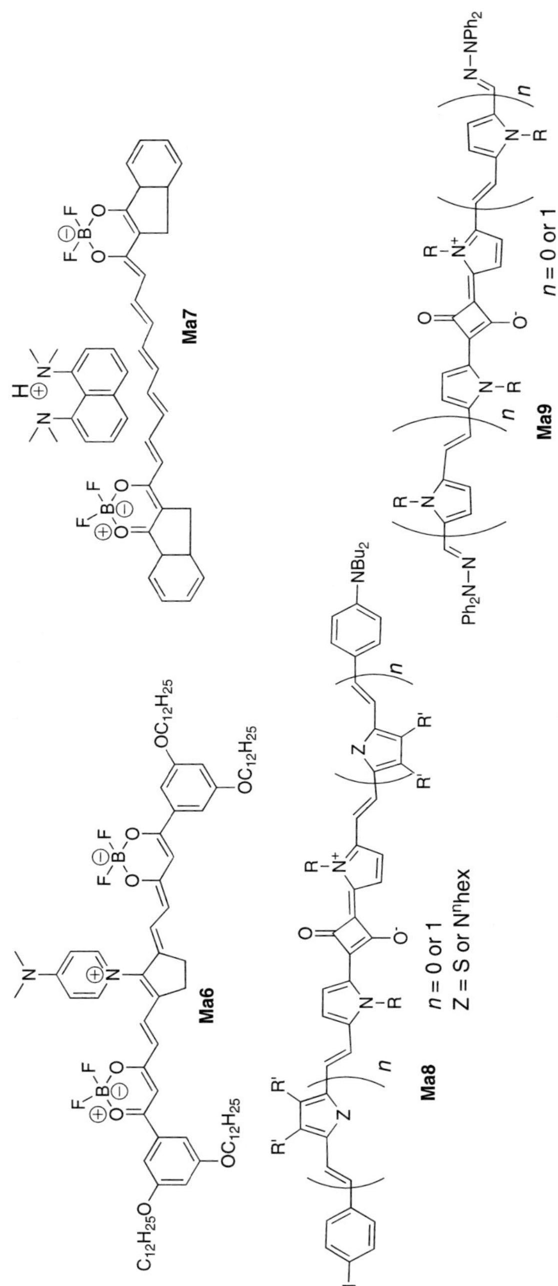

Fig. 18 Symmetric bisdioxaborine polymethines and squaraines **Ma6–Ma9** for TPA in the NIR

near IR, Michler's ketone derivatives **MK1** and **MK2** were considered (Fig. 17) with σ_{TPA} values up to 325 GM at 900 nm [102, 103].

As in the case of dipolar molecules, a new field of research appeared recently in IR, particularly at telecommunications wavelengths. The group of Marder and Perry developed bisdioxaborine polymethines **Ma6** and **Ma7** (Fig. 18) with large third-order nonlinearities in solution and solid state [104]. The same group designed conjugated squaraines **Ma8** and **Ma9** with vinylene groups and electron-rich heterocycles as bridges (Fig. 18) [95]. Large TPA cross-sections have been obtained (up to 33 000 GM around 1 μm; 1600 and 800 GM at 1.3 and 1.5 μm, respectively); these values have been ascribed to small detuning energy terms (0.32–0.39 eV) and high dipole transition moments μ_{01} (18.4–20.4 D).

Our group in Lyon developed near-IR symmetric cyanines **An2** and **An3** (Fig. 12) for TPA-based optical limiting [69]. Significant TPA cross-sections have been found for these molecules (544 and 731 GM at 1495 and 1437 nm, respectively); it is worth noting that, for this family, the dipolar analog **An1** presents a higher efficiency of 792 GM, while usually dipolar systems are less active than their symmetric parents. Excited state absorption, interesting for the optimization of optical limiting, was shown for these molecules between 1400–1600 nm.

In conclusion to this part, Kuzyk published several papers related to the existence of a fundamental limit for the cross-section σ_{TPA} of organic molecules [105, 106]. According to sum-over states rules, this maximum cross-section σ_{TPA}^{max} is given by the relationship Eq. 8 at resonance. In this equation, σ_{TPA}^{max} and energies E_{01} and E_{02} are expressed respectively GM and in eV; n is the refractive index at the incident laser energy and N the electrons number in the molecule.

$$\sigma_{TPA}^{max} = 63.5 \left(\frac{1}{n^2} \left(\frac{n^2 + 2}{3} \right)^2 \right) \left(\frac{E_{20}}{2E_{10} - E_{20}} \right)^2 \left(\frac{E_{20}}{\Gamma_{20}} \right) \left(\frac{N^2}{E_{10}^3} \right). \tag{8}$$

A relevant characteristic of the relation Eq. 8 is the quadratic proportionality between σ_{TPA}^{max} and N. In these papers, all considered organic molecules were shown to present σ_{TPA} values falling far below σ_{TPA}^{max}. Unexpectedly, molecules with the largest number of electrons feature a weak ratio $\sigma_{TPA}/\sigma_{TPA}^{max}$. This allowed Kuzyk to conclude that the model, relating TPA efficiency and molecular structure, and which led to the design of different families of molecules as described above, is not well optimized and does not take advantage of all electrons. Kuzyk suggests to consider the ratio σ_{TPA}/N^2 as the relevant parameter to evaluate the TPA efficiency of molecules.

In the following section, different types of branched molecules will be described and we are going to discuss the possibility to obtain an enhancement of TPA efficiency by inducing molecules interactions within considered TPA systems.

2.2
Branched Molecules

The branched molecules approach to create interactions between chromophores and enhance σ_{TPA} values has been widely used. Different systems have been designed; we will report some of them and the results for TPA response optimization. The factor $F_{(n)}$, defined in Eq. 9 as the ratio between the TPA cross-section $\sigma_{TPA}^{(n)}$ of the branched molecule with n branches of monomers and TPA cross-section $\sigma_{TPA}^{(1)}$ of the monomer, will be used to draw conclusions about interactions within the branched molecule: $F_{(n)} = n$ will be ascribed to independent branches, while $F_{(n)} \neq n$ will correspond to deleterious ($F_{(n)} < n$) or constructive ($F_{(n)} > n$) interactions between branches.

$$F_{(n)} = \frac{\sigma_{TPA}^{(n)}}{\sigma_{TPA}^{(1)}}. \qquad (9)$$

Zyss has widely demonstrated the interest of multipolar systems for quadratic NLO [107–109]; on the basis of his theory, efficient octupolar quadratic nonlinear systems have been designed [110–116]. This symmetry has been considered as a new strategy for the design of efficient systems in the field of TPA. In a preliminary theoretical study Brédas et al. predicted, in the case octupoles with three independent branches, an increase by a factor of 3 of σ_{TPA} values ($F_{(n)} = 3$) with respect to those of their dipolar analogs [117]; furthermore, this study suggests a larger enhancement, when the structuring core allows an electronic coupling between branches ($F_{(n)} > 3$).

2.2.1
Metallic Systems

Different octupolar metallic complexes have been studied. Theoretical papers have shown that σ_{TPA} values depend strongly on the nature of the ligand L, of the metal and of the symmetry of the complex [118, 119]. A large enhancement in σ_{TPA} values has been shown from dipolar complexes D to octahedral octupoles O, with intermediate values for tetrahedral octupoles T (Fig. 19).

Experimental studies correlate these theoretical studies. Le Bozec et al. designed a series of octahedral Ru(II), Ni(II), Cu(II) and Zn(II) tris(bipyridyl) LB complexes (Fig. 20) [120]. This work shows clearly that TPA properties can be strongly tuned by the nature of the metal, since σ_{TPA} values of 2200, 1900, 1700, 1050 and 650 GM at 765 nm were observed for Ru(II), Fe(II), Zn (II), Cu(II) and Ni(II), respectively. The role of the metal has been also demonstrated in 1,10-phenanthroline-based complexes [121], and indirectly from the comparison between data of Le Bozec et al. and those obtained by Coe et al. from Ru(II) and Fe(II) complexes with electron acceptor pyridinium groups (Fig. 20); weaker σ_{TPA} values were found for these complexes: 5, 62,

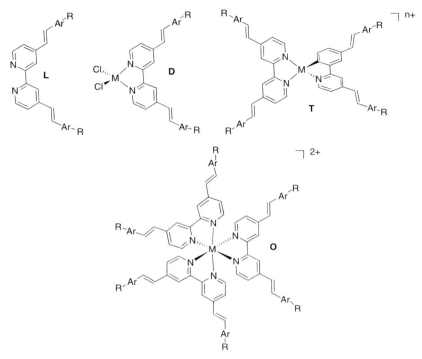

Fig. 19 Dipolar **D**, tetrahedral **T** and octupolar **O** metallic complexes based on the ligand **L**; n depends on the charge of the metal

120, 180 and 92 GM at 750 nm for **Co1–5**, respectively, while Fe(II) feature very low values (≤ 13 GM) [122]. The influence of the ligand on TPA properties has been demonstrated on 1,10-phenanthroline derivatives [123].

Although a strong increase was predicted in these systems by theoretical calculations [124], a recent paper demonstrated that Zn(II) does not produce an enhancement of the TPA efficiency with respect to that of the ligand in octupolar complexes of Fig. 21 [125]; both complexes **C1** and **C2** feature σ_{TPA} values (214 and 530 GM, respectively) close to 3-times those of ligands **Lg1** and **Lg2** (69 and 190 GM, respectively).

In the field of metallic multibranched molecules for TPA, Humphrey et al. developed several series of organo-metallic complexes (Fig. 22). The iron octupolar complex **Hu1b** exhibits a TPA cross-section of 920 GM at 695 nm, which corresponds to an enhanced value with respect to that of the linear system **Hu1a** (66 GM) in a ratio of 14 between TPA cross-sections of both structures (Table 4); it must be noted that this efficiency increase is obtained with a red shift of the linear absorption (460 against 436 nm, respectively, as reported in Table 4) [126].

Alkylruthenium complexes **Hu2** have been also designed for TPA (Fig. 22). Linear absorption properties show a weak bathochromic shift from the

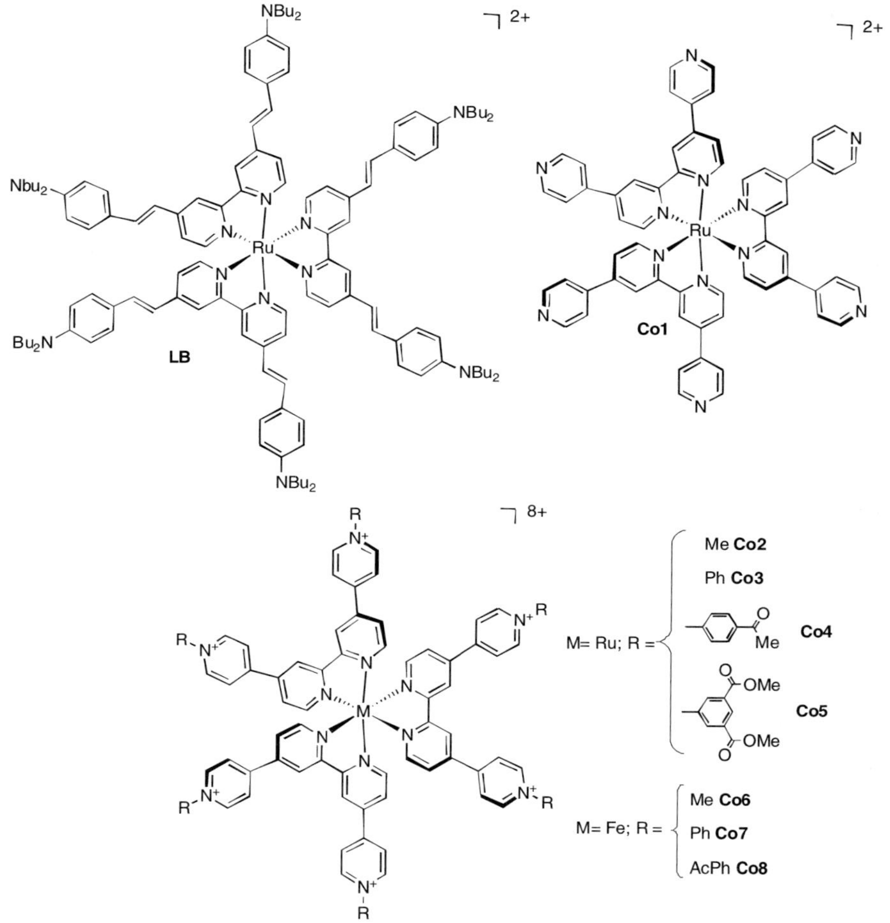

Fig. 20 Ruthenium(II) complexes **LB** and **Co1–Co8**; n depends on the charge of the metal

monomer to the octupole in non-nitro systems, with a higher transparency of the dendrimer **Hu2d**; a similar trend has been observed from the octupole in nitro-containing systems (**Hu2c′** and **Hu2d′**), with a significant red shift in the series with respect to the non-nitro containing analog. Concerning TPA properties, the dimer (octupole) **Hu2b** (**Hu2c**), with a value of $F_{(2)}$ ($F_{(3)}$) of 1.7 (2.3) at 800 nm, which corresponds to the MLCT (metal–ligand charge transfer) excitation band, gives rise to no (or a weak and deleterious) interaction (Table 4) [127–131]; the behavior of complexes **Hu2b′** and **Hu2c′** is similar. The dendrimer of second generation **Hu2d**, with a ratio $F_{(9)}$ of 15.5, exhibits significant interactions, while the existence of deleterious couplings in the derivative **Hu2d′**, for which a ratio $F_{(9)}$ of 7.3 has been found.

Table 4 TPA properties of multi-branched molecules; only fs data were reported

Molecules	λ_{abs}^{max} (nm)	λ_{TPA}^{meas} (nm)	n	$\sigma_{TPA}^{(n)}$ (GM)	$\sigma_{TPA}^{(1)}$ (GM)	$F_{(n)}$	Fig.	Ref.
Hu1a	436	695	1	66	66	1.0	22	60
Hu1b	460	695	3	920	66	14.0	22	60
Hu2a(a′)	382(386)	800	1	310	310(−)	1.0	22	61
Hu2b(b′)	407(463)	800	2	530(1200)	310	1.7	22	61
Hu2c(c′)	412(459)	800	3	700(1300)	310	2.3(2.2)	22	61
Hu2d(d′)	402(467)	800	9	4800(4400)	310	15.5(7.3)	22	61
Pr1	399	796	1	8.7	8.7	1.0	25	64
Pr2	417	796	2	27.4	8.7	3.1	25	64
Pr3	426	796	3	59.8	8.7	6.9	25	64
Ro1	392	800	1	4.6	4.6	1.0	26	68
Ro2	418	800	2	12.3	4.6	2.7	26	68
Ro3	429	800	3	15.8	4.6	3.4	26	68
Ro1′	377	800	1	3.6	3.6	1.0	26	68
Ro2′	386	800	2	7.8	3.6	2.2	26	68
Ro3′	393	800	3	12.2	3.6	3.4	26	68
Re1	389	670	1	320	320	1.0(2.0)	28	75
Re2	412	680	3	1300	320	4.1(8.1)	28	75
Re3	417	694	5	2700	320	8.4(16.9)	28	75
Re4	413	694	13	4500	320	14.1(28.1)	28	75
BD4	351	710	1	95	95	1.0	29	78
BD5	355	710	3	290	95	3.1	29	78
BD6	378	740	1	130	130	1.0	29	78
BD7	384	735	3	470	130	3.6	29	78
Pr4	402	740	1	94	94	1.0	30	79
Pr5	412	752	3	603	94	6.4	30	79
Pr6	413	770	6	1412	94	15.0	30	79

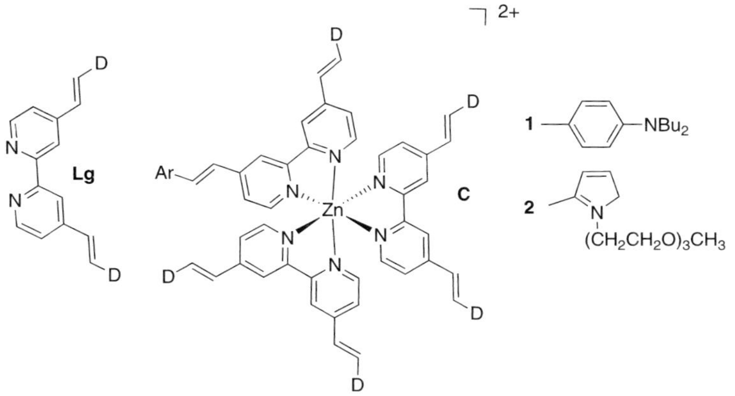

Fig. 21 Zn(II) complexes **C1** and **C2** and the parent ligands **Lg1** and **Lg2**

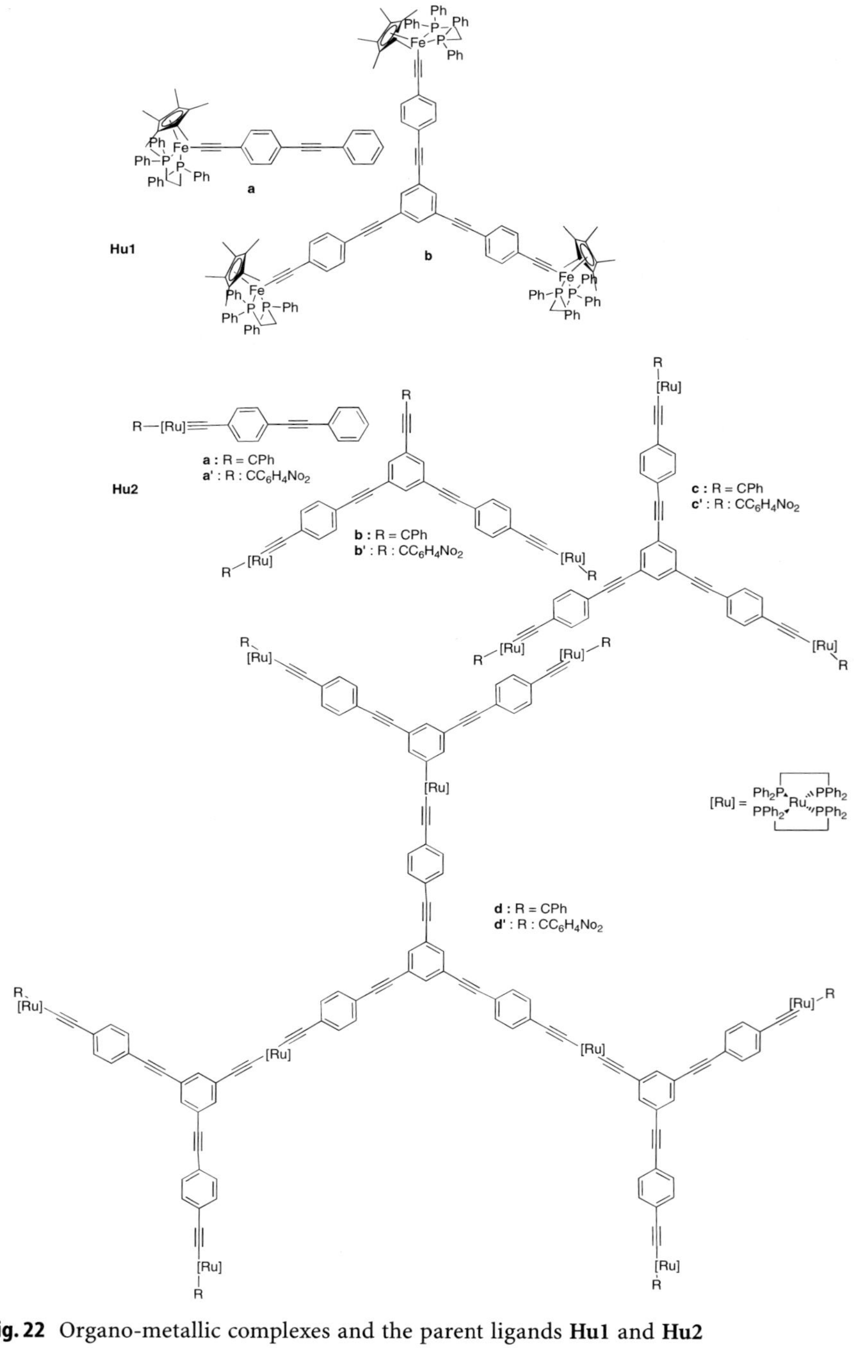

Fig. 22 Organo-metallic complexes and the parent ligands **Hu1** and **Hu2**

2.2.2
Organic Systems

Experimental values of σ_{TPA} obtained for crystal violet **CV** and brilliant green **BG** (Fig. 23) confirm the potentialities of octupolar molecules, for which TPA cross-sections of 1980 and 762 GM have been reported respectively by excitation in the state $2A'$; the efficiency of **CV** has been found comparable to that of dipolar or quadrupolar systems [117].

The group of Cheah designed zigzag oligoaryleethynylenes **C1–C5** (Fig. 24) with up to six dipolar units [132]. Photophysical properties of molecules **C1–C5** are summarized in Table 5. Two types of observations can be deduced

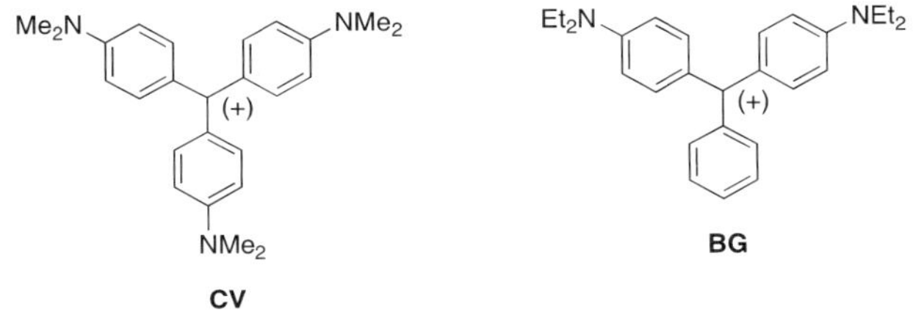

Fig. 23 Crystal violet **CV** and brilliant green molecules **BG**

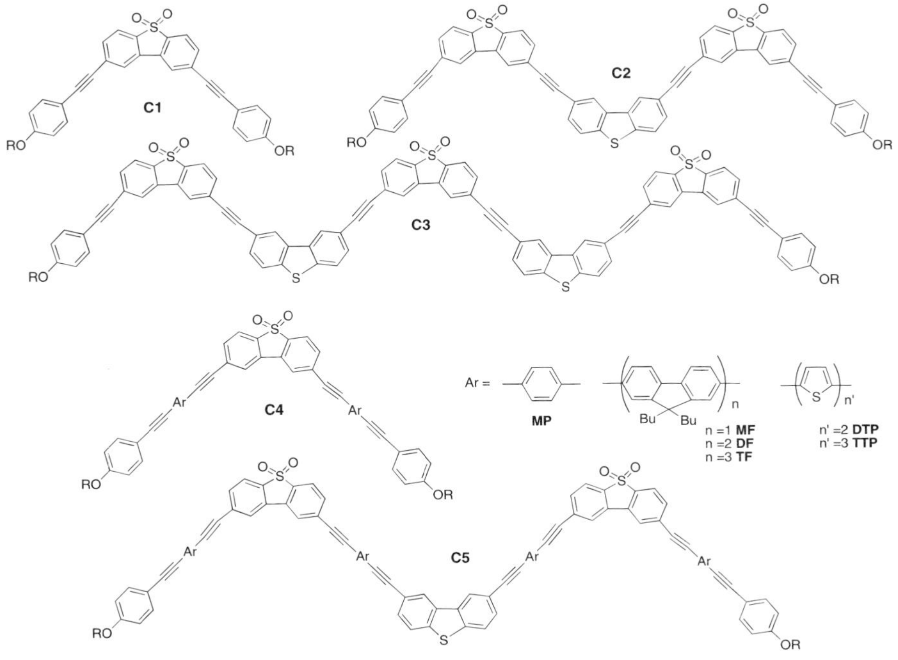

Fig. 24 Zigzag oligoaryleethynylenes molecules **C1–C5**

Table 5 Spectroscopic properties of molecules C1–C5

Molecules	λ_{abs}^{max} (nm)	λ_{em}^{max} (nm)	σ_{TPA}^{800} (GM)
C1	302, 341	440	–
C2	302, 341	443	–
C3	299, 341	446	–
C4(MP)	345	466	25
C5(MP)	345	466	23
C4(MF)	364	478	212
C5(MF)	367	476	433
C4(DF)	373	483	378
C5(DF)	374	485	723
C4(TF)	377	498	609
C5(TF)	378	499	1214
C4(DTP)	405	534	264
C4(TTP)	425	553	465

from this Table: (1) the increase of the conjugation length within each branch leads to a red shift of linear absorption and emission bands and to an increase of the σ_{TPA} value (comparison between **C4(MF)**, **C4(DF)** and **C4(TF)**, or **C5(MF)**, **C5(DF)** and **C5(TF)**, or **C4(DTP)** and **C4(TTP)**); (2) the increase of branches number within the same family does not lead to a significant shift of bands position (comparison between **C1**, **C2** and **C3**, or **C4(MP)** and **C5(MP)**, or **C4(MF)** and **C5(MF)**, or **C4(DF)** and **C5(DF)**, or **C4(TF)** and **C5(TF)**), while the value of the TPA cross-section for the molecule **C4** is half the value of the molecule **C5** for the same family (except for the series **MP**). This last trend clearly illustrates the lack of interaction between each dipolar unit within this type of system.

Many dendrimeric structures investigated for enhanced TPA properties were based on a triphenylamine core. Prasad et al. studied TPA properties of oxadiazole derivatives **Pr1–3** (Fig. 25) [133]. Bathochromic shifts in linear absorption (emission spectra) from 399 (503) to 417 (510) and 426 (516) nm in **Pr1** to **Pr2** and **Pr3** respectively indicate interactions between chromophores with electronic delocalization. This trend is confirmed by TPA cross-sections, for which $F_{(n)}$ takes values of 1.0, 3.1 and 6.9 from **Pr1**, **Pr2** and **Pr3**, respectively (Table 4). These interactions were ascribed, from ab initio calculations, to electronic and vibronic couplings [134]. Goodson III et al. contributed to the understanding of the NLO efficiency enhancement mechanism by the study of dynamics processes (vibrational distortions, energy redistribution, energy transfer, long-range interactions) in this type of multibranched structure [135]. This work led him to the conclusion that delocalization of the optical excitation over several branches in **Pr1**, **Pr2** and **Pr3**, involves a coherent energy transfer due to strong intramolecular coupling interactions [136].

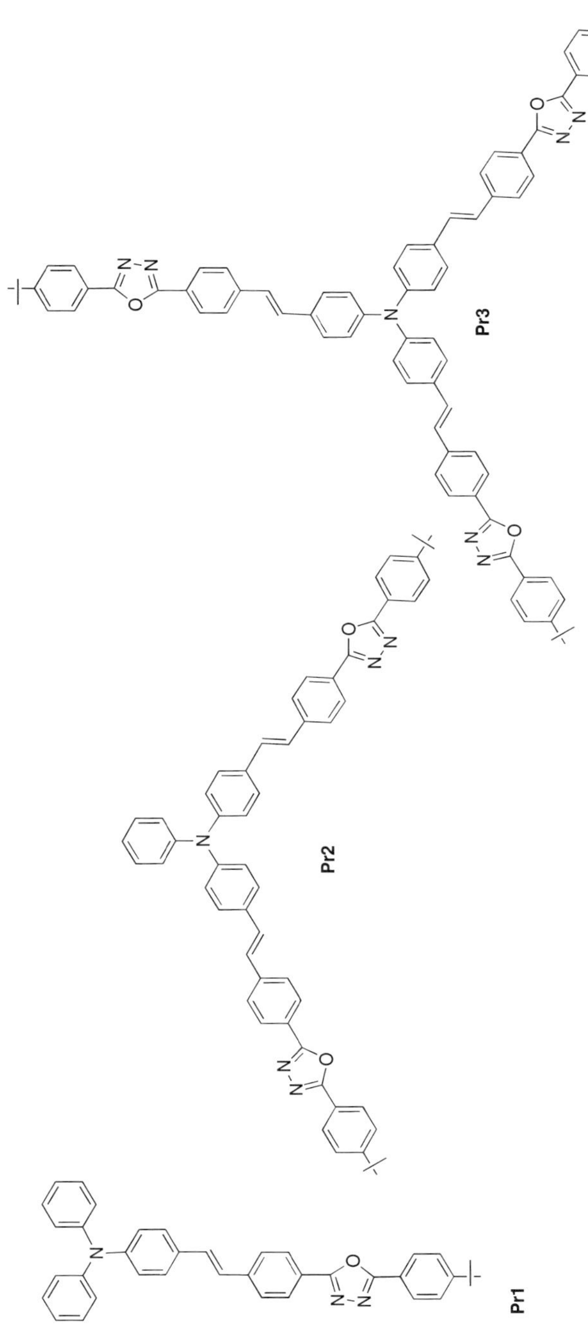

Fig. 25 Branched molecules **Pr1–Pr3**

Different other papers reported the comparison of TPA properties in linear, quadrupolar and octupolar analog systems with triphenylamine as the structuring core. Coupling between arms did not lead to TPA enhancements as high as those obtained by Prasad et al. Two series of benzothiazole derivatives **Ro1–3** and **Ro1′–3′**, which differ by an additional phenyl ring in each branch in the last series, have been designed (Fig. 26) [137]; a strong bathochromic shift has been obtained within each series, while the introduction of the phenyl ring led to a better transparency for systems with the same symmetry. TPA properties vary similarly for both series in the ratio 1.0; 2.7; 3.4 and 1.0; 2.2; 3.4 (Table 4). In spite of the weak interactions, these molecules were shown to exhibit high nonlinear transmission properties related to excited-state absorption from singlet and triplet excited levels. Systems bearing sulfonyl groups were designed (Fig. 27) [138]; their TPA efficiency has been found to vary in the ratio 1.0; 2.2; 3.2 and 1.0; 4.7; 11.1 depending on the TPA maximum. The wavelength dependence of this strong enhancement is illustrated in Fig. 27; calculations have shown a multidimensional intramolecular charge transfer from the donor triphenylamine to the peripheral groups. Other works on triphenylamine cored dimers and octupole TPA properties revealed a similar trend with ratios 1.0; 2.3; 3.1 [139], 1.0; 1.7; 4.2 and 1.0; 2.1; 3.5 [140], 1.0; 2.2; 2.8 [141] and 1.0; 2.5; 4.7 [142].

As the interest of these multibranched stilbene-type structures has been already theoretically demonstrated [143], Rebane et al. proposed dendrimers based on triphenylamine stilbene derivatives **Re1–Re4** (Fig. 28) [144, 145]. The TPA efficiency varies in the ratio 1.0; 4.1; 8.4; 14.1 from the parent **Re1** to the second generation dendrimer **Re4**, as reported in Table 4[4]. The authors choose the number of triphenylamines N or the number of π-electrons N_π to investigate size dependence of the TPA response of these systems; they show that σ_{TPA} appears to scale as (N^2) or (N_π^2) with a saturation for higher generations of dendrimers; a similar trend has been observed in the case of three-photon absorption. This was ascribed to an interbranch π-conjugation phenomenon; it was shown that **Re3**, with an effective maximum number of π-electrons of nearly 150, is the optimum dendrimer generation in terms of the largest coherence domain size. Similar dendrimers with SBu as peripheral groups have been shown to exhibit very high σ_{TPA} values (11 000 GM for the fourth generation dendrimer); however, σ_{TPA} varies linearly with the total number of stilbene moieties [146].

Corroborating above experimental data on triphenylamine cored dendrimers, a theoretical analysis of the TPA enhancement of these systems revealed that it is strongly dependent on the nature of the parent chromophore (dipolar or quadrupolar) [147]: the strong increase of TPA efficiency of dipoles in triphenylamine cored systems is interpreted in terms of polar-

---

[4] We only use data of papers 148 and 149 corresponding to systems bearing exactly the same substituents.

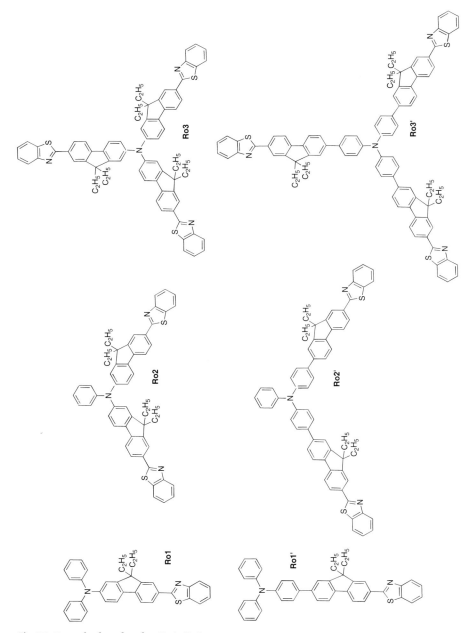

Fig. 26 Branched molecules **Ro1–Ro3**

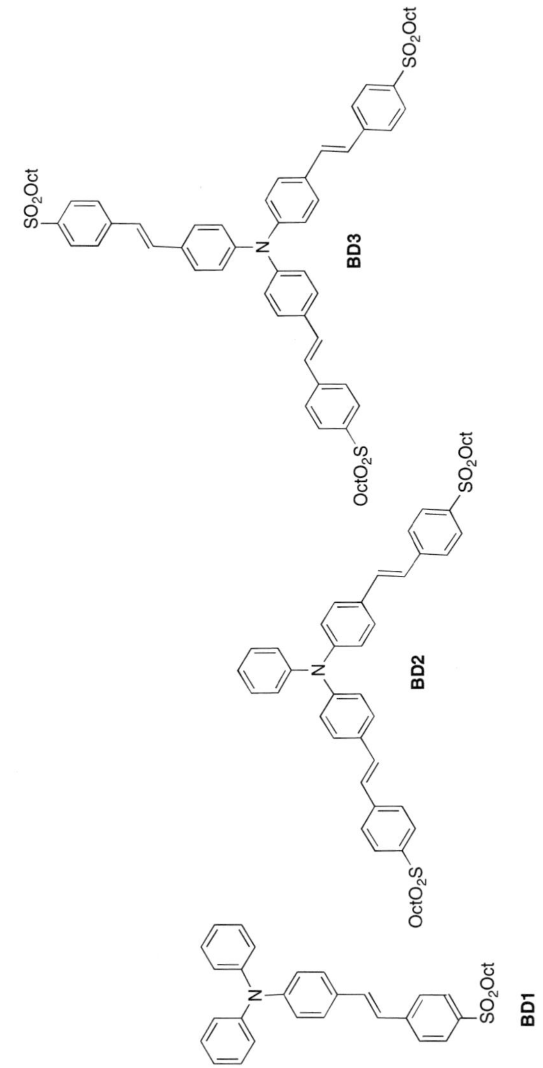

Fig. 27 Branched molecules BD1–BD3

ization enhancement in a system considered as a monomer bearing a more efficient electronic withdrawing group (due to the triphenylamine donor common to the three branches), while the weak enhancement in the case of quadrupoles, in spite of coherent interactions between branches, is due to the significant dissymmetry of quadrupolar branches. In contrast, in the case of triphenylbenzene octupoles **BD4** and **BD6**, based on dipolar monomers **BD5** and **BD7** (Fig. 29), the three branches are shown to behave independently

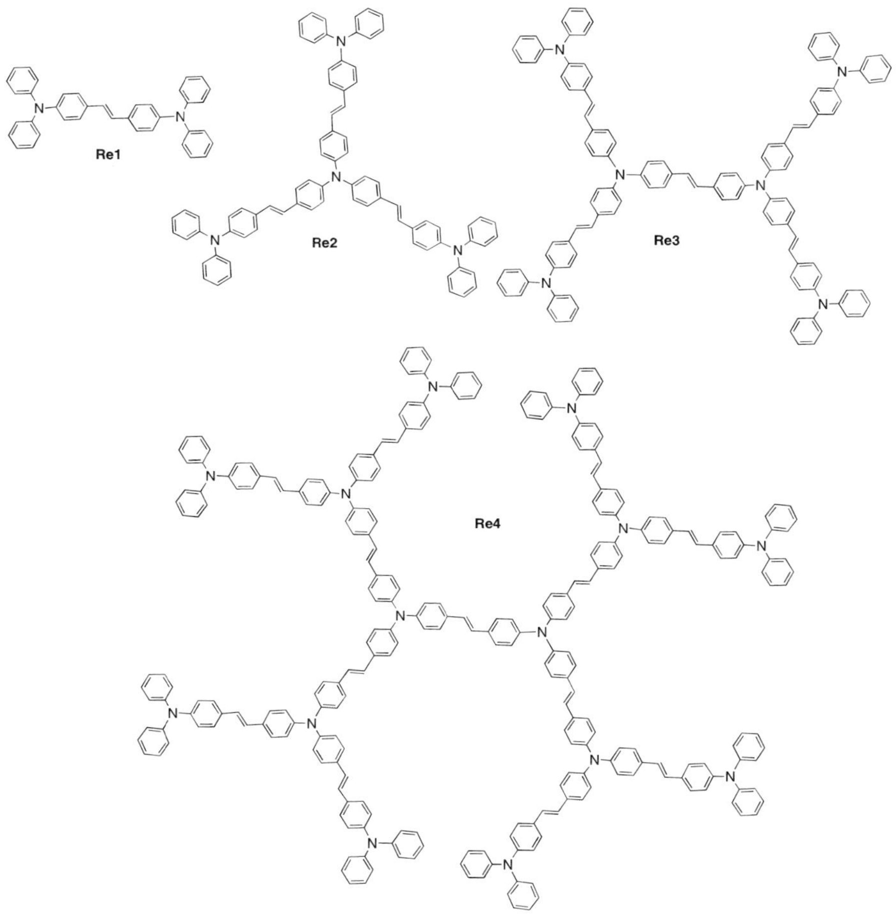

Fig. 28 Branched molecules **Re1–Re4**

(Table 4), due to only electrostatic interactions and not coherent couplings as in the case of the triphenlamine core [148].

The truxene core has been shown to be also very efficient to enhance TPA properties (Fig. 30) [149]; the σ_{TPA} value of **Pr5** is 6.4-times higher than that of **Pr4**, while **Pr6** presents a TPA efficiency 15.0-times larger than that of **Pr4** (Table 4). It is worth noting that this intramolecular extended π-conjugation in the octupole **Pr5** and the dendrimer **Pr6** leads to a two-photon excited fluorescence enhancement of 190% and 135% respectively with respect to that of the monomer **Pr4**.

Finally, the concept of the [2.2]paracyclophane, which has been introduced by Zyss and Bazan et al. as an out of plane core inducing through-space delo-

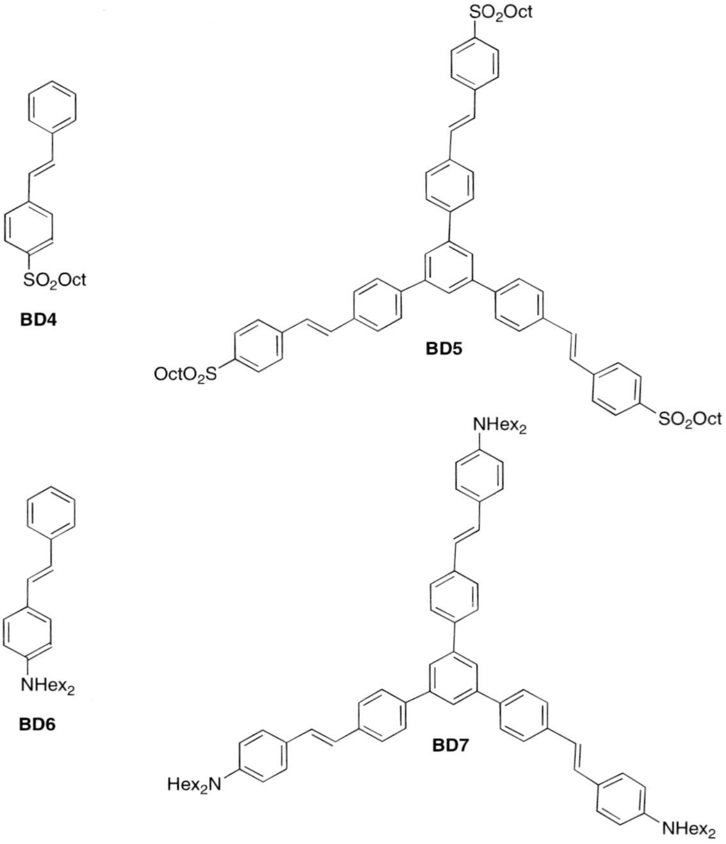

Fig. 29 Branched molecules BD4–BD7

calization within octupolar systems [150], has been used in the TPA field with arms of varying length (Fig. 31) [151].

The linear absorption spectra are red shifted, when increasing the length of the molecule in linear systems from **Ba3L** to **Ba5L**, as in paracyclophanes dimers from **Ba3P** to **Ba5P** (see values reported in Fig. 31); it is worth noting that shoulders in absorption and emission spectra of dimers **Ba3P** and **Ba5P**, ascribed to a "Davydov-like splitting", have been observed. As far as TPA data is concerned, no effect of the dimerization has been observed: in good agreement with theoretical calculations, no change in TPA maxima was observed from the linear to the corresponding dimer, as reported in Fig. 31; furthermore, the additive effect of the dimerization on σ_{TPA} values can be concluded from $F_{(2)}$ values, which have been calculated to be 1.6 and 2.4 in the case of **Ba3** and **Ba5**, respectively.

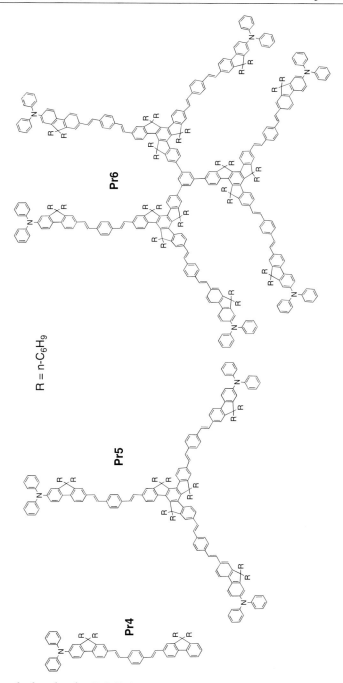

Fig. 30 Branched molecules **Pr4–Pr6**

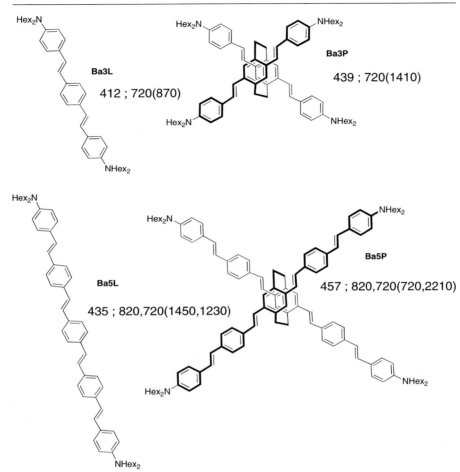

Fig. 31 [2.2]Paracyclophane branched molecules **Ba3P** and **Ba5P**, with the corresponding linear branches **Ba3L** and **Ba5L**. The wavelengths λ_{max}; $\lambda_{TPA}(\sigma_{TPA})$ are expressed in nm, σ_{TPA} in GM

The following conclusion from all examples of families reported above can be expressed: although some dendrimer families have been reported to present nonlinear enhancements in terms of TPA response, in particular dendrimers based on triphenylamines, most branched molecules were shown to exhibit no perturbation, i.e. only additive effect, in their TPA efficiency with respect to that of the corresponding linear system; it is worth noting that several other dendritic systems have been shown to confirm this trend [152]. As for linear molecules, Kuzyk presents the limits of this dendrimeric approach [105, 106].

3
Enhancement of Two-Photon Absorption by Excitonic Coupling

An alternative approach, no longer based on the existence of charge transfer but related to excitonic coupling interactions between monomers within oligomers and dendrimers, was developed by our group. It is a promising approach for obtaining enhanced σ_{TPA} values in the visible, for which the best compromise transparency/nonlinearity is a real challenge.

The possibility to obtain the enhancement of $\chi^{(3)}$ nonlinear susceptibilities by excitonic coupling in molecular aggregates was theoretically proposed by Spano et al. in 1989 [153]. Their analysis was based on interacting two-level systems. They found that the $\chi^{(3)}$ of small aggregates contain terms that scale as N^2 and $N(N-1)$ that could lead to giant nonlinearities. Strong effects of excitonic coupling in the two-photon absorption spectrum of the DEANST crystal have been reported by Feneyrou et al. [154]. The DEANST crystal has a broad absorption band that is rationalized using the framework of Frenkel–Davydov exciton theory, and does not display the strong molecular TPA resonance that is due to the internal charge-transfer transition. The two-photon absorption spectrum of organic nanocrystals grown in gel glasses has been reported by Sanz et al. [155]. However, it did not show any clear TPA enhancement.

The TPA properties of polyene oligomers were investigated in the mid 1990s, and a strong correlation between their TPA and their lengths has been established. The oligomer properties of thiophene, furan and pyrrole where calculated by Ågren et al. (Fig. 32) [156]. In these oligomers, σ_{TPA} was shown to increase as $L^{7.0}$, $L^{6.2}$, and $L^{4.0}$ respectively, with L the head to tail length of the molecule.

In 1998, our group pointed out that dimers and small oligomers of phenyls, fluorene and stilbene have enhanced TPA cross-sections with their size. This alternative strategy to design efficient two-photon sensitive molecules without grafting donor/acceptor substituent groups on a conjugated core led to a better trade-off between the TPA strength and the linear absorption cut-off for optical limiting applications at visible wavelengths. The experimental data and CNDO/S calculations were rationalized with Davydov's excitonic model of interacting monomers.

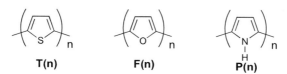

Fig. 32 Oligomers T(n), F(n) and P(n)

3.1
CNDO/S Calculation of Enhanced Two-Photon Absorption with Polyphenyl Oligomers

Typical CNDO/S linear and two-photon absorption spectra of polyphenyls are displayed in Fig. 33 [157]. The low energy optical response of all compounds is accurately described by a three-level model with parameters that are summarized in Table 6. Transition energies are red-shifted, and transition dipole moments increase with the monomer number n.

The TPA cross-sections σ_{TPA} of polyphenyls ($n = 1$ to 6, with n the number of benzene rings) are displayed in Table 6 with the TPA resonance energies. These calculations show a large increase of σ_{TPA} as a function of n according to a power law:

$$\sigma_{TPA} = 1.1 \times 10^{-49} n^{3.0} .$$

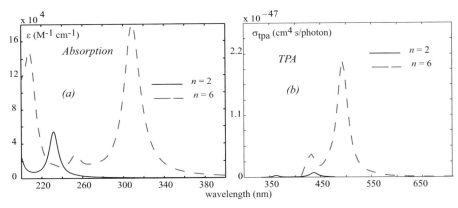

Fig. 33 Absorption spectra of biphenyl and sexiphenyl. (**a**) Linear absorption; (**b**) two-photon absorption

Table 6 Excited states parameters of polyphenyls from CNDO/S calculations

n	E_{01} (cm^{-1})	E_{02} (cm^{-1})	μ_{01} (D)	μ_{12} (D)	σ_{TPA} (10^{-50} cm^4 s/photon)
2	43 103	45 662	6.0	9.0	80
3	37 594	42 373	8.5	11.4	330
4	35 461	42 017	9.9	12.0	610
5	33 670	41 152	11.6	16.2	1260
6	32 362	40 486	12.6	18.3	2090

Table 7 Power dependency α of parameters describing the two-photon absorption of polyphenyls in the three-level model

P	μ_{01}	μ_{02}	$\left(E_{01} - \frac{E_{02}}{2}\right)^{-1}$
α	0.67	0.64	0.47

No saturation effect was predicted up to $n = 6$ in good agreement with experimental data on the tetra and the pentaphenyl. The increase of σ_{TPA} values with n occurs with a red shift of the first absorption and of the σ_{TPA} bands.

The variations of each parameter P describing the two-photon absorption of polyphenyls in the three-level model can be described by a power law. The corresponding powers α are displayed in Table 7.

Using these exponents in the three-level model, one finds a σ_{TPA} power dependency $\alpha = 3.5$ in good agreement with the power $\alpha = 3.0$ obtained from the CNDO/S calculation. Thus, on the basis of the three-level model, the enhancement of polyphenyl TPA can be rationalized by the simultaneous increase of ground-state transition dipole moment μ_{01}, excited-state transition dipole moment μ_{12}, and energy resonances with the number n of monomer units. The power dependency of transition dipole moments, i.e. $\alpha = 0.67$ and 0.64, is comparable with the power $\alpha = 0.5$ obtained in the exciton model for linear aggregates [158]. The difference in power laws is due to the conjugation that occurs between monomers.

3.2
Perturbative Three-Level Excitonic Model for Linear Oligomers

In this section, we present a simple perturbative model to describe the one- and two-photon absorption properties of linear oligomers [159]. We demonstrate that these dipole–dipole interactions determine the size dependency of optical properties, and in particular the enhancement of oligomer TPA.

A centro-symmetric monomer with only three eigenstates: $|0\rangle$, $|1\rangle$ and $|2\rangle$ is considered. The one-photon transition and two-photon transition are

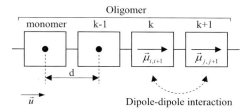

Fig. 34 Model oligomeric structure

$|0\rangle \to |1\rangle$ and $|0\rangle \to |2\rangle$, respectively. An oligomer consisting of N identical monomers is considered (Fig. 34).

This oligomer is assumed linear along $\vec{u}$, a unitary vector. Each next monomer is assumed to be the image of the previous one through translation along $d\vec{u}$. It is also assumed that $\vec{\mu}_{01} = \mu_{01}\vec{u}$ and $\vec{\mu}_{12} = \mu_{12}\vec{u}$. Monomers are assumed interacting via excitonic coupling, what we considered here as a dipole–dipole interaction between the transition electric dipole moments of monomers. Since this interaction may not be pure, correction factors are introduced that lead to an "effective" excitonic coupling. Dipole–dipole interactions are assumed limited to nearest neighbors, which means that a monomer k only interacts with monomers $k - 1$ and $k + 1$. The energy resulting from the interaction between the transition dipole moments $\vec{\mu}_{i_k,(i+1)_k}$ and $\vec{\mu}_{j_{k+1},(j+1)_{k+1}}$ is assumed equal to the following corrected classical dipole–dipole interaction energy:

$$V_{i,j} = - e_1 W \tau_{i,i+1}^{j,j+1} \frac{\mu_{i,i+1}\mu_{j,j+1}}{\mu_{01}^2}, \tag{10}$$

with $W = 2\mu_{01}^2/(e_1 4\pi\varepsilon_0 d^3)$ and where $\tau_{i,i+1}^{j,j+1}$ represents the previously mentioned correction factors.

We here assume that V_{ij} matrix elements are much smaller than $e1$ and $e2 - e1$. The perturbative treatment is applied up to the first order for wave functions and up to the second order for energies. The initial unperturbed oligomer is described with $\{|i_1 i_2 ... i_N\rangle\}$ as a basis set, in which i_k means that the monomer number k is in the state $|i\rangle$. For convenience, elements 0_k are suppressed from these representations; for example $|1_1 0_2 ... 0_N\rangle$, which means that all monomers are in state $|0\rangle$ except the first one which is in state $|1\rangle$, is simply written $|1_1\rangle$. Here, we are only concerned with the changes induced by oligomerization on the one- and two-photon absorptions initially exhibited by the monomer. The perturbative treatment is then limited to states $|0\rangle$, $|1\rangle$ and $|2\rangle$. The fundamental state $|0\rangle$ leads to a new fundamental state called $|\tilde{0}\rangle$. The N first degenerate excited states $|1_k\rangle$ and the N second degenerate excited states $|2_k\rangle$ lead respectively to N new excited states called $|\tilde{1}_p\rangle$ and to N new excited states called $|\tilde{2}_p\rangle$ with p varying from 1 to N. The corresponding wavenumbers are:

$$\bar{\omega}_{\tilde{0}\tilde{1}_p} = \frac{e_1}{hc}\left[1 - 2\tau_{01}^{01} W \cos\left(\frac{p\pi}{N+1}\right) \right.$$
$$- \frac{\eta^2 W^2 \left\{(\tau_{01}^{12})^2 + (\tau_{12}^{01})^2\right\}}{\xi}\left\{1 - \frac{2}{N+1}\sin^2\left(\frac{p\pi}{N+1}\right)\right\}$$
$$\left. - \frac{W^2 (\tau_{01}^{01})^2 (N-1)}{2}\left\{\underbrace{1 - \frac{4}{N+1}\sin^2\left(\frac{p\pi}{N+1}\right) - 1}_{N>2}\right\}\right], \tag{11}$$

$$\bar{\omega}_{\tilde{0}\tilde{2}_p}/2 = \frac{e_2}{2hc}\left\{1 - \frac{2\eta^2 W^2}{\xi(2-\xi)}\left[\frac{\left(\tau_{01}^{12}\right)^2 + \left(\tau_{12}^{01}\right)^2}{2}\left(1 - \frac{2}{N+1}\sin^2\left(\frac{p\pi}{N+1}\right)\right)\right.\right.$$

$$\left.+\,\tau_{01}^{12}\tau_{12}^{01}\cos\left(\frac{p\pi}{N+1}\right)\right]$$

$$\left.-\frac{\left(\tau_{01}^{01}\right)^2 W^2}{2\xi}\underbrace{\left[N - 3 + \frac{4}{N+1}\sin^2\left(\frac{p\pi}{N+1}\right) - N + 1\right]}_{N>2}\right\}, \quad (12)$$

where $\bar{\omega}_{XY}$ represents the wavenumber associated to the energy difference between states $|X\rangle$ and $|Y\rangle$. It is also easily shown that the electric dipole moments between those new states are all null except $\langle\tilde{0}|\vec{\mu}|\tilde{1}_p\rangle$ and $\langle\tilde{1}_p|\vec{\mu}|\tilde{2}_{p'}\rangle$ for any p and p'. This means that, from state $|\tilde{0}\rangle$, one-photon absorptions only lead to states $|\tilde{1}_p\rangle$ and two-photon absorptions to state $|\tilde{2}_p\rangle$.

A molecular material consisting of a homogeneous and isotropic dispersion of monodisperse oligomers in a transparent homogenous, linear and isotropic matrix is considered. τ_{01}^{12} and τ_{12}^{01} are assumed equal. The oscillator strengths $f_{\tilde{0}\tilde{1}_p}$ and the two-photon absorption strengths $f_{\tilde{0}\tilde{2}_p}^{TPA}$ associated to the one- and two-photon absorptions of interest have then the following expressions:

$$\frac{f_{\tilde{0}\tilde{1}_p}}{Nf_{01}} = \frac{2}{N(N+1)}\frac{1-(-1)^P}{2\tan^2\left(\frac{p\pi}{2(N+1)}\right)}, \quad (13)$$

$$\frac{f_{\tilde{0}\tilde{2}_p}^{TPA}}{Nf_{02}^{TPA}} = \frac{2}{N(N+1)}\frac{1-(-1)^P}{2\tan^2\left(\frac{p\pi}{2(N+1)}\right)}$$

$$\times\left\{1 + \frac{2W\tau_{01}^{01}}{2-\xi}\left[\left(6 - \xi + \frac{2\tau_{01}^{12}}{\tau_{01}^{01}}\right)\cos\left(\frac{p\pi}{N+1}\right)\right.\right.$$

$$\left.\left.+\frac{4\tau_{01}^{12}}{\tau_{01}^{01}\xi}\left(1 - \frac{2}{N+1}\sin^2\left(\frac{p\pi}{N+1}\right)\right)\right]\right\}. \quad (14)$$

A short mathematical analysis of formulae Eq. 13 and Eq. 14 shows that the absorptions corresponding to $p = 1$ and $p' = 1$ are highly dominating the one- and two-photon absorption spectra.

3.3
Two-Photon Absorption Properties of 9,9-Dihexylfluorene Oligomers

In this section, the previous model is used for analyzing the one- and two-photon absorption properties of the 9,9'-dihexylfluorene oligomers (Fig. 35) [161]. Their one- and two-photon absorption experimental data are

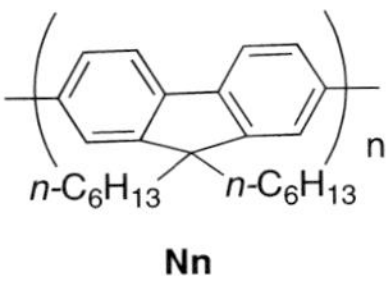

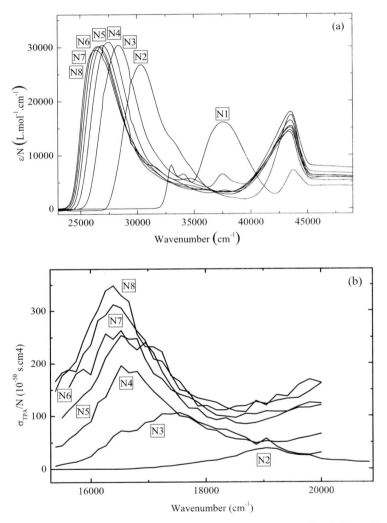

Fig. 35 9,9′-Dihexylfluorene oligomers

Fig. 36 Experimental one- and two-photon absorption spectra of 9,9′-dihexylfluorene oligomers in dichloromethane. *Top*: Absorption coefficient per monomer. *Bottom*: Two-photon absorption cross-section per monomer. Nx indicates an oligomer containing x monomeric units

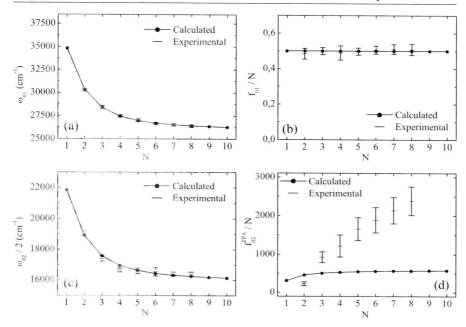

Fig. 37 Comparison between experimental and calculated optical properties of 9,9'-dihexylfluorene oligomers. *Top*: One-photon absorption. *Bottom*: Two-photon absorption. *Left*: Absorption energies. *Top right*: Oscillator strength per monomer. *Bottom right*: Two-photon absorption strength per monomer. Experimental uncertainties are displayed

reported in Fig. 36. As can be seen, an enhancement of two photon absorption per monomer is effectively obtained when the oligomer size increases [161].

The agreement between experimental and calculated data is very good, except for two-photon absorption strengths (Fig. 37).

Calculated data are obtained by varying the six effective parameters of the model e_1, e_2, μ_{01}, μ_{02}, τ_{01}^{01} and τ_{01}^{12}, and by fitting the corresponding predicted data with the 28 experimental data corresponding to oligomers with sizes 2 to 8. The complete model Hamiltonian corresponding to the fitted parameters has been diagonalized and the resulting transition energies, oscillator strengths and two-photon absorption strengths were calculated to test the relevance of the perturbative treatment. The overall behavior of energies does not change drastically when the Hamiltonian is diagonalized. Similarly, the behavior of oscillator strengths is only slightly affected. On the contrary, two-photon absorption strengths are changed towards a better agreement with experimental data, particularly in case of small τ_{12}^{12} values. It can then be concluded that the quantitative disagreement between experimental and calculated two-photon absorption strengths takes one of its main origins in the use of a perturbative treatment.

3.4
Two-Photon Absorption Properties of Fluorene V-Shapes and Dendrimeric Structures

Particular examples of V-shape and dendrimeric structures are represented in Fig. 38. A V-shape molecule is noted Vn and consists of two identical 9,9-dihexylfluorene oligomers of size n, substituted in meta position on a benzene core. A dendrimer is noted DnGg and consists of identical 9,9-dihexylfluorene oligomers of size n organized around benzene cores, on which they are substituted in *meta* position. Each benzene core bears three oligomers and g represents the dendritic generation.

Bifluorene derivative properties have been described previously [162]. Two-photon absorption properties are displayed in Fig. 39 for all branched molecules. Whatever the way monomers are organized, an increase of their number globally implies a decrease of both one- and two-photon absorption energies as well as an increase of the two-photon absorption strength per monomer, while the oscillator strength remains mainly constant. It shows however simultaneously that optical properties may strongly vary depending on the way the constituting monomers are organized. D2G1 and N6, for example, both consist of six monomers. N6 is however a better two-photon absorber than D2G1, with a two-photon absorption strength (1889) which is however 3.3-times larger than that of D2G1 (571), whereas D2G1 exhibits a better one-photon transparency, since its one-photon absorption energy (28 653 cm^{-1}) is 1986 cm^{-1} higher than that of N6 (26 667 cm^{-1}).

Thus, the coupling and spatial structuring of centrosymmetric units lead to an enhancement of two-photon absorption per monomeric unit. A figure-of-merit of all compounds investigated in this study is displayed in Fig. 40. High nonlinear absorption properties in the visible ascribed to these excitonic couplings have been described for this dendritic family [163].

3.5
Partial Diagonalization of the Three-Level Excitonic Interaction Hamiltonian Operator

In this section, we go beyond the perturbative treatment and calculate the eigenstates of large excitonic systems by diagonalizing the Hamiltonian operator on an efficient reduced basis set [164].

The full basis set of D1G3, which is here the largest investigated molecule with 21 monomeric units, contains around 10^{10} ($= 3^{21}$) vectors and the corresponding Hamiltonian then exhibits 10^{20} matrix elements. This makes the diagonalization process rather tricky and time consuming, even with iterative diagonalization methods. We are interested in the changes induced via the oligomerization and dendrimerization processes on both low-lying one- and two-photon absorptions of the investigated compounds. These latter initially

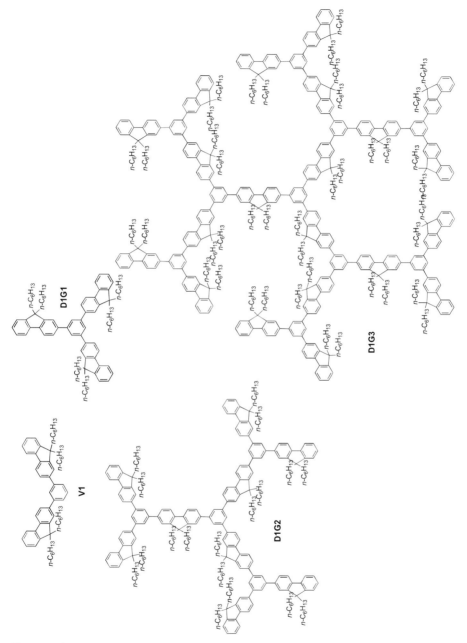

Fig. 38 Schematic representation of branched monofluorene derivatives: V-shape compound **V1** , dendrimeric compounds **D1G1**, **D1G2** and **D1G3**

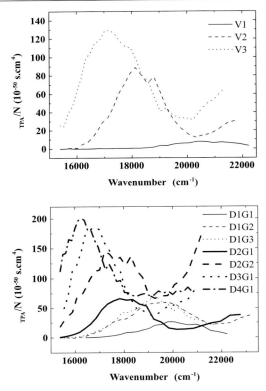

Fig. 39 Two-photon absorption spectra of V-shape and dendrimeric compounds. σ_{TPA}/N represents the two-photon absorption cross-section per monomer

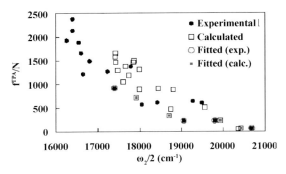

Fig. 40 Figure of merit of all investigated compounds using both experimental and calculated data

correspond to both $|0\rangle \rightarrow |1\rangle$ and $|0\rangle \rightarrow |2\rangle$ transitions of the monomeric unit. The three lowest excited states of the molecular system, which correspond to the ground state and to the excitation of one single monomer over the whole oligomeric or dendrimeric structure, exhibit a direct interaction with excited states that correspond to up to a maximum of three excited monomeric units. We then chose to restrict the diagonalization to the space defined by the basis vectors corresponding to a maximum of up to three excitations. This strongly reduces the size of the basis set, which, for example, now only contains 11 523 vectors for D1G3. The reduced basis sets of N2, N3, V1 and D1G1, which are the compounds used for fitting the parameters of the excitonic model, and which contain a maximum of three monomeric units, are the full basis sets of these systems.

Since the reduced basis sets remain quite large, and since we are only interested in the low-lying one- and two-photon absorptions of each molecular system, a full diagonalization of the Hamiltonian matrix is not necessary. We then use the Davidson iterative algorithm [165] for performing this operation, except for N2, N3, V1 and D1G1, where a full diagonalization is performed [166]. One- and- two-photon absorption properties are calculated using the $2N + 1$ lowest lying eigenvectors. For each one of them, the one- and two-photon absorption energies, the oscillator strength, and the two-photon absorption strength are calculated.

We use the smallest possible number of reference compounds for fitting the parameters of the excitonic model. The underlying idea is to validate the ability of the model to provide access to the expected properties of large structures from the knowledge of the experimental data of the smallest ones. First, e_1, e_2, μ_{01}, and μ_{12}, as well as F–F junction parameters τ_{01}^{01}, τ_{01}^{12}, and τ_{12}^{12}, are fitted on N2 and N3 data. Second, V- and D-type F–P–F junction parameters τ_{01}^{01}, τ_{01}^{12}, and τ_{12}^{12} are fitted on V1 and D1G1 data, respectively.

The comparison between calculated and experimental data is graphically represented in Fig. 41. It can immediately be seen, even if the number of data used for the fitting procedure (16 for 13 parameters) is smaller than in our perturbative approach used for linear oligomers (28 for 6 parameters), that the agreement between experimental and calculated data is much better.

The agreement is better for small structures (high absorption energies and low oscillator strength and two-photon absorption strength) than for large structures. This is not very surprising, first, because the splitting of the initially noninteracting $|1\rangle$ and $|2\rangle$ states of the monomeric units constituting the molecular structure becomes larger when the size of the system increases, which necessarily involves a higher contribution of the highly excited states that are not included in the reduced basis set, and, second, because the three-level approximation becomes less adequate when the size of the system increases. The second main deviation concerns the oscillator strengths. It seems indeed that this property remains theoretically constant over all molecular structure, whereas it experimentally varies. The conservation of the oscillator strength per monomer is the result of the well known Thomas–Reiche–Kuhn sum rule [167–170]. It is then intrinsic to the model and independent of the fitted parameters. Experimental molecular structures however involve phenyl centers that are energetically included in our model via the τ_{ab}^{cd} junc-

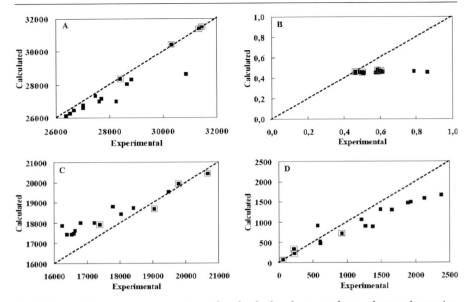

Fig. 41 Correlation between experimental and calculated one- and two-photon absorption properties of oligomeric, V-shape and dendrimeric structures. (**A**) One-photon absorption energy (cm^{-1}). (**B**) Oscillator strength per monomer. (**C**) Two-photon absorption energy (cm^{-1}). (**D**) Two-photon absorption strength per monomer

tion parameters, but that are not included in overall transition electric dipole moments.

4
Conclusion

In the last ten years, many works have designed efficient two-photon absorbers in relation with fast-spreading applications such as fluorescence bioimaging, optical limiting, and microfabrication. The first part of this work summarizes the state of the art of these researches. The main strategy has been to design quadrupolar molecules with symmetrical electron donor or acceptor substituents. World record-breaking molecules have been reported with two-photon cross-sections reaching the tens of thousand GM. Such molecules are highly conjugated with high dipole transition moments (< 10 D). They are colored and their two-photon resonances occur at near infrared laser wavelengths. However, studies on fundamental limits of molecular two-photon cross-sections indicate that these molecules are far from reaching the limits, i.e. they do not take advantage of all electrons. More recently, numerous dendrimeric molecules based on dipolar branches have been designed to obtain larger two-photon cross-sections. Spectroscopic studies have pointed out the strong dipolar interactions that occurs between branches, and the following delocalization of optical excitation over several branches. If some dendrimer families present some nonlinearity enhancement, most branched molecules have only additive nonlinearity. This is probably due to the charge transfer term $(\mu_{01}^2 \Delta \mu_{01}^2)$ of two-photon cross-sections that may be only additive.

In the second part of this work we review our theoretical and experimental works to obtain enhanced two-photon cross-sections by using the superlinear response of centrosymmetric monomers that are coherently coupled. In this alternative approach, the nonlinear material consists of an assembly of nonsubstituted π-electron systems that are coupled by dipole–dipole interactions. The monomer two-photon term is a pure transition dipole term $(\mu_{01}^2 \mu_{12}^2)$. Typical materials can be molecular aggregates, nanocrystals, oligomers, and dendrimers. The dipole–dipole interactions determine the size dependency of optical properties, and in particular of two-photon cross-sections.

A perturbative excitonic model based on N interacting three-level systems is presented for linear oligomers of size N. Analytical expressions for one-photon and two-photon energies and absorption strengths are derived. Despite the splitting in N excited states, the one- and two-photon absorptions are dominated by the lowest excited states, and oligomer optical responses can be approximated by new three-level systems. Transition energies are redshifted with a dependency on coupling energy and oligomer size that is similar to previous results on interacting two-level systems. The one-photon oscillator strength increases linearly with the oligomer size, while the two-photon absorption strength has a superlinear dependency that depends on the coupling energy and the oligomer size. There is a good agreement between calculated and experimental data of fluorene oligomers, except for two-photon absorption strengths that are underestimated by this perturbative treatment.

Enhancement of two-photon cross-sections by two-dimensional and three-dimensional arrangements of monomers has been demonstrated with fluorene V-shapes and dendrimeric structures. Such multidimensional structures lead to lower two-photon absorptions than linear oligomers, but they have better one-photon transparencies. An accurate calculation of large exitonic systems is obtained by diagonalizing the Hamiltonian operator on a reduced basis set.

This theoretical and experimental work on model fluorene oligomeric and dendritic structures opens the way for an alternative strategy to design two-photon absorbers without the need for donor and acceptor substituents. This new approach may lead to a better trade-off between high nonlinearities and transparency in the visible. More importantly, it substantiates the need for more two-photon studies on molecular nanocrystals that can provide a strong packing of interacting monomers, and therefore giant nonlinearities.

References

1. Göppert-Mayer M (1931) Ann Phys 9:273
2. Reinhardt BA (1999) Photon Sci News 4:21
3. He GS, Bhawalkar JD, Zhao CF, Prasad PN (1995) Appl Phys Lett 67:2433
4. Ehrlich JB, Wu XL, Lee IYS, Hu ZY, Röckel H, Marder SR, Perry JW (1997) Opt Lett 22:1843
5. Silly MG, Porrès L, Mongin O, Chollet PA, Blanchard-Desce M (2003) Chem Phys Lett 379:74
6. Maruo S, Nakamura O, Kawata S (1997) Opt Lett 22:132
7. Joshi MP, Pudavar HE, Swiatkiewicz J, Prasad PN, Reinhardt BA (1999) Appl Phys Lett 74:170
8. Cumpston BH, Ananthavel SP, Barlow S, Dyer DL, Ehrlich JE, Erskine LL, Heikal AA, Kuebler SM, Lee IYS, McCord-Maughon D, Qin J, Röckel H, Rumi M, Wu XL, Marder SR, Perry JW (1999) Nature 398:51
9. Kawata S, Sun HB, Tanaka T, Takada K (2001) Nature 412:697
10. Zhou W, Kuebler SM, Braum KL, Yu T, Cammack JK, Ober CK, Perry JW, Marder SR (2002) Science 296:1106
11. Wang I, Bouriau M, Baldeck PL, Martineau C, Andraud C (2002) Opt Lett 27:1348
12. Sun HB, Takada K, Zaccaria RP, Kim MS, Lee KS, Kawata S (2003) Appl Phys Lett 83:1104
13. Sun HB, Suwa T, Takada K, Zaccaria RP, Kim MS, Lee KS, Kawata S (2004) Appl Phys Lett 85:3708
14. Klein S, Barsella A, Leblond H, Bulou H, Fort A, Andraud C, Lemercier G, Mulatier JC, Dorkenoo K (2005) Appl Phys Lett 86:211118(1)
15. Lin CL, Wang I, Pierre M, Colombier I, Andraud C and Baldeck PL (2005) Nonlinear Opt Quantum Opt 14:375
16. Bhawalkar JD, He GS, Park CK, Zhao CF, Ruland G, Prasad PN (1996) Opt Commun 124:33
17. He GS, Helgeson R, Lin TC, Zheng Q, Wudl F, Prasad PN (2003) IEEE J Quantum Electron 39:1003
18. He GS, Lin TC, Hsiao VKS, Cartwright KN, Prasad PN, Natarajan LV, Tondiglia VP, Jakubiak R, Vaia RA, Bunning TJ (2003) Appl Phys Lett 83:2733
19. Bhawalkar JD, Kumar ND, Zhao CF, Prasad PN (1997) J Clin Laser Med Surg 15:201
20. Dichtel WR, Serin JM, Edder C, Fréchet JMJ, Matuszewski M, Tan LS, Ohulchanskyy TY, Prasad PN (2004) J Am Chem Soc 126:5380
21. Oar MA, Serin JM, Dichtel WR, Fréchet JMJ, Ohulchanskyy TY, Prasad PN (2005) Chem Mater 17:2267
22. Atif M, Dyer PE, Paget TA, Snelling HV, Stringer MR (2007) Photodiag Photodyn Ther 4:106
23. Parthenopoulos DA, Rentzepis PM (1989) Science 245:843
24. Strickler JH, Webb WW (1993) Adv Mater 5:479
25. Pudavar HE, Joshi MP, Prasad PN, Reinhardt BA (1999) Appl Phys Lett 74:1338
26. Day D, Gu M, Smallridge A (2001) Adv Mater 13:1005
27. Belfield KD, Schafer KJ (2002) Chem Mater 14:3656
28. Belfield KD, Liu Y, Negres RA, Fan M, Pan G, Hagan DJ, Hernandez FE (2002) Chem Mater 14:3663
29. Denk W, Strickler JH, Webb WW (1997) Science 248:73
30. Gura T (1990) Science 276:1988

31. Wang X, Krebs LJ, Al-Nuri M, Pudavar HE, Ghosal S, Liebow C, Nagy AA, Schally AV, Prasad PN (1999) Proc Natl Acad Sci USA 96:11 081
32. Wang X, Pudavar HE, Kapoor R, Krebs LJ, Bergey EJ, Liebow C, Prasad PN, Nagy A, Schally AV (2001) J Biomed Opt 6:319
33. Larson DR, Zipfel WR, Williams RM, Clark SW, Bruchez MP, Wise FW, Webb WW (2003) Science 300:1434
34. Swiatkiewicz J, Prasad PN, Reinhardt BA (1998) Opt Commun 157:135
35. Albota M, Beljonne D, Brédas JL, Ehrich JE, Fu JY, Heikal AA, Hess SE, Kogej T, Levin MD, Marder SR, MacCord-Maughon D, Perry JW, Röckel H, Rumi M, Subbramaniam G, Webb WW, Wu XL, Xu C (1998) Science 281:1653
36. Orr BJ, Ward JF (1971) Mol Phys 20:513
37. Meyers F, Marder SR, Pierce BM, Brédas JL (1994) J Am Chem Soc 116:10703
38. Andraud C, Anémian R, Collet A, Nunzi JM, Morel Y, Baldeck PL (2000) J Opt A Pure Appl Opt 2:284
39. Oudar JL, Chemla DS (1977) J Chem Phys 66:2664
40. Bhawalkar D, He GS, Park CK, Zhao CF, Ruland G, Prasad PN (1996) Opt Commun 124:33
41. He GS, Yuan L, Cheng N, Bhawalkar JD, Prasad PN, Brott LL, Clarson SJ, Reinhardt BA (1997) J Opt Soc Am B 14:1079
42. He GS, Yuan L, Prasad PN, Abotto A, Facchetti A, Pagani GA (1997) Opt Commun 140:49
43. Beljonne D, Kogej T, Marder SR, Perry JW, Bredas JL (1999) Nonlinear Opt 21:461
44. Luo Y, Norman P, Macak P, Ågren H (2000) J Phys Chem 104:4718
45. Pati SK, Marks TJ, Ratner MA (2001) J Am Chem Soc 123:7287
46. Zalesny R, Bartowiak W, Styrcz S, Leszczynski J (2002) J Phys Chem A 106:4032
47. Lei H, Huang ZL, Wang HZ, Tang XJ, Wu LZ, Zhou GY, Wang D, Tian YB (2002) Chem Phys Lett 352:240
48. Zhou X, Ren AM, Feng JK, Liu XJK (2003) J Phys Chem A 107:1850
49. Day PN, Nguyen KA, Patcher R (2006) J Chem Phys 125:094103
50. Song YZ, Li DM, Song XN, Huang XM, Wang CK (2006) J Mol Struct (THEOCHEM) 772:75
51. Ohta K, Antonov L, Yamada S, Kamada K (2007) J Chem Phys 127:084 504
52. Reinhardt BA, Brott LL, Clarson SJ, Dillard AG, Bhatt JC, Kannan R, Yuan L, He GS, Prasad PN (1998) Chem Mater 10:1863
53. Delysse S, Raimond P, Nunzi JM (1997) Chem Phys 219:341
54. Kogej T, Beljonne D, Meyers F, Perry JW, Marder SR, Brédas JL (1998) Chem Phys Lett 298:1
55. Barzoukas M, Blanchard-Desce M (2000) J Chem Phys 113:3951
56. Terenziani F, Le Droumaguet C, Mongin O, Katan C, Blanchard-Desce M (2006) Nonlin Opt Quantum Opt 35:69
57. Strehmel B, Sarker AM, Detert H (2003) Chem Phys Chem 4:249
58. Marder SR, Torruellas WE, Blanchard-Desce M, Ricci V, Stegeman GI, Gilmour S, Brédas JL, Li J, Bublitz GU, Boxer SG (1997) Science 276:1233
59. De Boni L, Misoguti L, Zilio SC, Mendonça CR (2005) Chem Phys Chem 6:1121
60. Hales JM, Hagan DJ, Van Stryland EW, Schafer KJ, Morales AR, Belfield KD, Pacher P, Kwon O, Zojer E, Brédas JL (2004) J Chem Phys 121:3152
61. Belfield KD, Morales AR, Kang BS, Hales JM, Hagan DJ, Van Stryland EW, Chapel VM, Percino J (2004) Chem Mater 16:4634
62. Liu ZQ, Fang Q, Wang D, Cao DX, Xue G, Yu WT, Lei H (2003) Chem Eur J 9:5074

63. Abbotto A, Beverina L, Bozio R, Bradamante S, Ferrante C, Pagani GA, Signorini R (2000) Adv Mater 12:1963

64. Kannan R, He GS, Yuan L, Xu F, Prasad PN, Dombroskie AG, Reinhardt BA, Baur JW, Vaia RA, Tan LS (2001) Chem Mater 13:1896

65. Kim OK, Lee KS, Woo HY, Kim KS, He GS, Swiatkiewicz J, Prasad PN (2000) Chem Mater 12:284

66. Barsu C, Fortrie R, Nowicka K, Baldeck PL, Vial JC, Barsella A, Fort A, Hissler M, Bretonnière Y, Maury O, Andraud C (2006) Chem Commun p 4744

67. Picot A, Malvolti F, Le Guennic B, Baldeck PL, Williams JAG, Andraud C, Maury O (2007) Inorg Chem 46:2659

68. Beverina L, Fu J, Leclercq A, Zojer E, Pacher P, Barlow S, Van Stryland EW, Hagan DJ, Brédas JL, Marder SR (2005) J Am Chem Soc 127:7282

69. Bouit PA, Wetzel G, Berginc G, Loiseaux B, Toupet L, Feneyrou P, Bretonnière Y, Kamada K, Maury O, Andraud C (2007) Chem Mater 19:5325

70. Albota M, Beljonne D, Brédas JL, Ehrlich JE, Fu JY, Heikal AA, Hess SE, Kogej T, Levin MD, Marder SR, McCord-Maughon D, Perry JW, Röckel H, Rumi M, Subramaniam G, Webb WW, Wu XL, Xu C (1998) Science 281:1653

71. Rumi M, Ehrlich JE, Heikal AA, Perry JW, Barlow S, Hu Z, McCord-Maughon D, Parker TC, Röckel H, Thayumanavan S, Marder SR, Beljonne D, Brédas JL (2000) J Am Chem Soc 122:9500

72. Lee KS, Choi SW, Woo HY, Moon KJ, Shim HK, Jeong MY, Lim TK (1998) J Opt Soc Am B 15:393

73. Ventelon L, Moreaux L, Mertz J, Blanchard-Desce M (1999) Chem Commun p 2055

74. Lee WH, Cho M, Jeon SJ, Cho BR (2000) J Phys Chem A 104:11033

75. Iwase Y, Kamada K, Ohta K, Kondo K (2003) J Mater Chem 13:1575

76. Lee KS, Kim MS, Yang HK, Sun HB, Kawata S, Fleitz P (2004) Mol Cryst Liq Cryst 424:35

77. Lu Y, Hasegawa F, Goto T, Ohkuma S, Fukuhara S, Kawazu Y, Totani K, Yamashita T, Watanabe T (2004) J Luminesc 110:1

78. Masunov A, Tretiak A (2004) J Phys Chem 108:899

79. Lee KS, Yang WJ, Choi JJ, Kim CH, Jeon SJ, Cho BR (2005) Org Lett 7:323

80. Kish R, Nakano M, Yamada S, Kamada K, Ohta K, Nitta T, Yamaguchi K (2005) Synth Met 154:181

81. Yang G, Qin C, Su Z, Dong S (2005) J Mol Struct (THEOCHEM) 726:61

82. Zeng Q, He GS, Prasad PN (2005) J Mater Chem 15:579

83. Ohta K, Kamada K (2006) J Chem Phys 124:124303

84. Tao LM, Guo YH, Huang XM, Wang CK (2006) Chem Phys Lett 425:10

85. Zhao L, Yang G, Su Z, Qin C, Yang S (2006) Synth Met 156:1218

86. Ma W, Wu Y, Gu D, Gan F (2006) J Mol Struct (THEOCHEM) 772:81

87. Han DM, Feng JK, Ren AM, Zhang XB, Sun CC (2007) J Mol Struct (THEOCHEM) 802:67

88. Zhao Y, Xiao L, Wu F, Fang X (2007) Opt Mater 29:1206

89. Lee W-H, Cho M, Jeon S-J, Cho BR (2000) J Phys Chem 104:11033

90. Anémian R, Morel Y, Baldeck PL, Paci B, Nunzi JM, Andraud C (2003) J Mater Chem 13:2157

91. Paci B, Nunzi JM, Anémian R, Andraud C, Collet A, Morel Y, Baldeck PL (2000) J Opt A: Pure Appl Opt 2:268

92. Zheng S, Beverina L, Barlow S, Zojer E; Fu J, Padiha LA, Fink C, Kwon O, Yi Y, Shuai Z, Van Stryland EW, Hagan DJ, Brédas JL, Brédas SR (2007) Chem Commun p 1372

93. Drobizhev M, Stepanenko Y, Dzenis Y, Karotki A, Rebane A, Taylor PN, Anderson HL (2004) J Am Chem Soc 126:15352
94. Chung SJ, Rumi M, Alain V, Barlow S, Perry JW, Marder SR (2005) J Am Chem Soc 127:10844s
95. Chung SJ, Zeng S, Odani T, Beverina L, Fu J, Padiha LA, Biesso A, Hales JM, Zhan X, Schmidt K, Ye A, Zojer E, Barlow S, Hagan DJ, Van Stryland EW, Yi Y, Shuai Z, Pagani GA, Brédas JL, Perry JW, Marder SR (2006) J Am Chem Soc 186:14444
96. Zhang XB, Feng JK, Ren AM, Sun CC (2006) J Phys Chem 110:12222
97. Hayek A, Nicoud J-F, Bourgogne FC, Baldeck PL (2006) Angew Chem Int Ed 45:6466–6469
98. Ventelon L, Charier S, Moreaux L, Mertz J, Blanchard-Desce M (2001) Angew Chem Int Ed 40:2098
99. Mongin O, Porrès L, Charlot M, Katan C, Blanchard-Desce M (2007) Chem Eur J 13:1481
100. Martineau C, Anémian R, Andraud C, Wang I, Bouriau M, Baldeck PL (2002) Chem Phys Lett 362:291
101. Wang I, Bouriau M, Baldeck PL, Martineau C, Andraud C (2002) Opt Lett 27:1348
102. Martineau C, Lemercier G, Andraud C, Wang I, Bouriau M, Baldeck PL (2003) Synth Meth 138:353
103. Lemercier G, Martineau C, Mulatier JC, Wang I, Stéphan O, Baldeck PL, Andraud C (2006) New J Chem 30:1606
104. Hales JM, Zheng S, Barlow S, Marder SR, Perry JW (2006) J Am Chem Soc 128:11362
105. Kuzyk MG (2003) J Chem Phys 119:8327
106. Moreno JP, Kuzyk MG (2005) J Chem Phys 123:194101
107. Zyss J (1993) J Chem Phys 98:6583
108. Zyss J, Ledoux I (1994) Chem Rev 94:77
109. Weibel JD, Yaron D, Zyss J (2003) J Chem Phys 119:11847
110. Thalladi VR, Brasselet S, Weiss HC, Blaser D, Katz AK, Carrell HL, Boese R, Zyss J, Nangia A, Desiraju GR (1998) J Am Chem Soc 120:2563
111. Andraud C, Zabulon T, Collet A, Zyss J (1999) Chem Phys 245:243
112. Le Floc'h V, Brasselet S, Zyss J, Cho BR, Lee SH, Jeon SJ, Cho M, Min KS, Suh MP (2005) Adv Mater 17:196
113. Fiorini C, Charra F, Nunzi JM, Samuel IDW, Zyss J (1995) Opt Lett 20:2469
114. Brasselet S, Zyss J (1998) J Opt Soc Am B 15:257
115. Viau L, Bidault S, Maury O, Brasselet S, Ledoux I, Zyss J, Ishow E, Nakatani K, Le Bozec H (2004) J Am Chem Soc 126:8386
116. Piron R, Brasselet S, Josse D, Zyss J, Viscardi G, Barolo C (2005) J Opt Soc Am B 22:1276
117. Beljonne D, Wenseleers W, Zojer E, Shuai Z, Vogel H, Pond SJK, Perry JW, Marder SR, Brédas JL (2002) Adv Funct Mater 12:631
118. Zhang XB, Feng JK, Ren AM (2007) J Organomet Chem 692:3778
119. Zhang XB, Feng JK, Ren AM (2007) J Chem Phys 111:1328
120. Feuvrie C, Maury O, Le Bozec H, Ledoux I, Morrall JP, Dalton GT, Samoc M, Humphrey MG (2007) J Phys Chem A 111:8980
121. Girardot C, Lemercier G, Mulatier JC, Chauvin J, Baldeck PL, Andraud C (2007) Dalton Trans p 3421
122. Coe BJ, Samoc M, Samoc A, Zhu L, Yi Y, Shuai Z (2007) J Phys Chem A 111:472
123. Girardot C, Lemercier G, Mulatier JC, Andraud C, Chauvin J, Baldeck PL (2008) Tetrahedron Lett 49:1753
124. Liu XJ, Feng JK, Ren AM, Cheng H, Zhou H (2004) J Chem Phys 120:11493

125. Mazzucato S, Fortunati I, Scolaro S, Zerbetto M, Ferrante C, Signorini R, Pedron D, Bozio R, Locatelli D, Righetto S, Roberto D, Ugo R, Abbotto A, Archetti G, Beverina L, Ghezzi S (2007) Phys Chem Chem Phys 9:2999
126. Cifuentes MP, Humphrey MG, Morrall JP, Samoc M, Paul F, Lapinte C, Roisnel T (2005) Organometallics 24:4280
127. McDonagh AM, Humphrey MG, Samoc M, Luther-Davis B (1999) Organometallics 18:5195
128. McDonagh AM, Humphrey MG, Samoc M, Luther-Davis B, Houbrechts S, Wada T, Sasabe H, Persoons A (1999) J Am Chem Soc 121:1405
129. Powell CE, Morrall JP, Ward SA, Cifuentes MP, Notaras EGA, Samoc M, Humphrey MG (2004) J Am Chem Soc 126:12234
130. Cifuentes MP, Powell CE, Morrall JP, McDonagh AM, Lucas NT, Humphrey MG, Samoc M, Houbrechts S, Asselberghs I, Clays K, Persoons A, Isoshima T (2006) J Am Chem Soc 128:10819
131. Samoc M, Morrall JP, Dalton GT, Cifuentes MP, Humphrey MG (2007) Angew Chem Int Ed 46:731
132. Lo PW, Li KF, Wong MS, Cheah KW (2007) J Org Chem 72:6672
133. Chung SJ, Kim KS, Lin TC, He GS, Swiatkiewicz J, Prasad PN (1999) J Phys Chem B 103:10741
134. Macak P, Luo Y, Norman P, Ågren H (2000) J Chem Phys 113:7055
135. Goodson III TG (2005) Acc Chem 38:99
136. Wang Y, He GS, Prasad PN, Godson III TG (2005) J Am Chem Soc 127:10128
137. Rogers JE, Slagle JE, Mclean DG, Sutherland RL, Brant MC, Heinrichs J, Jakubiak R, Kannan R, Tan LS, Fleitz PA (2007) J Phys Chem A 111:1899
138. Katan C, Terenziani F, Mongin O, Werts HV, Porrès L, Pons T, Mertz J, Tretiak S, Blanchard-Desce M (2005) J Phys Chem A 109:3024
139. Zhang BJ, Jeon SJ (2003) Chem Phys Lett 377:210
140. Lee HJ, Sohn J, Hwang J, Park SY (2004) Chem Mater 16:456
141. Cui YZ, Fang Q, Huang ZI, Xue G, Xu GB, Yu WT (2004) J Mater Chem 14:2443
142. Wang X, Yang P, Xu G, Jiang W, Yang T (2005) Synt Met 155:464
143. Zhou X, Ren AM, Feng JK, Liu XJ (2002) Chem Phys Lett 362:541
144. Drobizhev M, Karotki A, Dzenis Y, Rebane A, Suo Z, Spangler CW (2003) J Phys Chem B 107:7540
145. Drobizhev M, Rebane A, Suo Z, Spangler CW (2005) J Luminesc 111:291
146. Drobizhev M, Karotki A, Dzenis Y, Rebane A, Spangler CW (2001) Opt Lett 26:1081
147. Katan C, Tretiak S, Werts MHV, Bain AJ, Marsh RJ, Leonczek N, Nicolau N, Badaeva E, Mongin O, Blanchard-Desce M (2007) J Phys Chem A 111:9468
148. Terenziani F, Le Droumaguet C, Katan C, Mongin O, Blanchard-Desce M (2007) Chem Phys Chem 8:723
149. Zheng Q, He GS, Prasad PN (2005) Chem Mater 17:6004
150. Zyss J, Ledoux I, Volkov S, Chernyak V, Mukamel S, Bartholomew GP, Bazan GC (2000) J Am Chem Soc 122:11956
151. Bartholomew GP, Rumi M, Pond SJK, Perry JW, Tretiak S, Bazan GC (2004) J Am Chem Soc 126:11529
152. Adronov A, Fréchet JMJ, He GS, Kim KS, Chung SJ, Swiatkiewicz J, Prasad PN (2002) Chem Mater 12:2838
153. Spano FC, Mukamel S (1989) Phys Rev A 40:5783
154. Feneyrou P, Baldeck PL (2000) J Phys Chem A 104:4764
155. Sanz N, Ibanez A, Morel Y, Baldeck PL (2001) Appl Phys Lett 78:2569
156. Norman P, Luo Y, Ågren H (1999) Opt Commun 168:297

157. Anémian R, Baldeck PL, Andraud C (2002) Mol Cryst Liq Cryst 374:335
158. McRae EG, Kasha M (1958) J Chem Phys 28:721
159. Fortrie R, Anémian R, Stephan O, Mulatier JC, Baldeck PL, Andraud C, Chermette H (2007) J Phys Chem C 111:2270
160. Anémian R, Mulatier JC, Andraud C, Stéphan O, Vial JC (2002) Chem Commun p 1608
161. Lemercier G, Mulatier JC, Martineau C, Anémian R, Andraud C, Wang I, Stéphan O, Amari N, Baldeck PL (2005) CR Chimie 8:1308
162. Barsu C, Anémian R, Andraud C, Stephan O, Baldeck PL (2006) Mol Cryst Liq Cryst 446:175
163. Barsu C, Andraud C, Amari N, Baldeck PL (2005) Nonlinear Opt Quantum Opt 14:311
164. Fortrie R, Barsu C, Baldeck PL, Andraud C, Chermette H (2008) J Phys Chem C, to be published
165. Davidson ER (1975) J Comput Phys 17:87
166. Mathematica 5.2, Wolfram Research Inc 2005
167. Kuhn W (1925) Z Phys 33:408
168. Thomas W (1925) Naturwissenschaften 13:627
169. Reiche F, Thomas W (1925) Z Phys 34:510
170. Heisenberg W (1925) Z Phys 33:879

Subject Index

Printing: Krips bv, Meppel, The Netherlands
Binding: Stürtz, Würzburg, Germany